Northern Spirits

A social history of alcohol in Canada

Other Books by Reginald G. Smart

The Smart Report: Substance Abuse and Canadian Youth
1993

The Steel Drug: Cocaine in Perspective
(with P. Erickson, E. M. Adlaf, G. F. Murray)
1987, 1994

Research Advances in Alcohol and Drug Problems, Volumes 1-9
(edited with H. Cappell, F. Glaser, Y. Israel, H. Kalant, E. Sellers, W. Schmidt)
1974-1986

Forbidden Highs: The Nature, Treatment and Prevention of Illicit Drug Abuse
1983

The New Drinkers: Teenage Use and Abuse of Alcohol
1976, 1980

Drug Education: Current Issues, Future Directions
(with D. Fejer)
1974

Social Class and the Treatment of Alcoholism: An Investigation of Social Class as a Determinant of Diagnosis, Prognosis and Therapy
(with W. Schmidt and M. Moss)
1968

Lysergic Acid Diethylamide (LSD) in the Treatment of Alcoholism
(with T. Storm, E. Baker and L. Solursh)
1967

Northern Spirits

A social history of alcohol in Canada

by Reginald G. Smart &
Alan C. Ogborne

ARF

Canadian Cataloguing in Publication Data

Smart, Reginald G., 1936-
Northern Spirits

2nd ed.
Includes bibliographical references.
ISBN 0-88868-276-X

1. Drinking of alcoholic beverages - Canada.
2. Alcoholism - Canada. I. Ogborne, Alan C. (Alan Charles), 1943- . II. Ontario.
Addiction Research Foundation. III. Title.

HV5306.S62 1996 362.29'2'0971 C96-931676-3

Addiction Research Foundation
33 Russell Street
Toronto, Ontario
M5S 2S1

Printed and bound in Canada.

NORTHERN SPIRITS:
A SOCIAL HISTORY OF ALCOHOL IN CANADA

Introduction

by Robin Room

In this volume, two of Canada's most respected alcohol scholars give us something rare — a fully rounded perspective on the story of alcohol in Canadian society, past and present. The story, it turns out, intertwines with many significant features of the general story of Canada over the years. In the trading between aboriginal Canadians and Europeans in early colonial Canada, it was fur which passed in one direction but often alcohol in the other. Later in the colonial period, it was from a tavern that the 1837 rebellion in Upper Canada was launched. After Confederation, the differences in political culture between anglophone and francophone Canada were dramatized by the results of the national plebiscite on Prohibition in 1898. More recently, the complex relations of Canada with its southern neighbor are exemplified and illuminated by the history of alcohol as a border issue — not only during Prohibition on both sides of the border, but also in the recent trade disputes sometimes known as the "beer wars" (Ferris, Room and Giesbrecht, 1993). The story, indeed, illustrates the dictum that "alcohol is to social science what dye is to microscopy" (Duster, 1983) — it provides a stain with which we can see much about the structure and social processes of Canadian society.

Now, as in the past, drinking is associated with sociability and good times for many Canadians. But drinking frequently brings problems. A pattern of heavy drinking can ruin the drinker's health, through such diseases as cirrhosis and cancer. The drinking may become an ingrained habit, from which it is difficult to recover. Single drinking occasions can also bring problems, including injury, disability and death.

And the problems from drinking happen not only to the drinker. The casualties from drinking and driving often include other drivers, passengers and pedestrians. Productivity in the workplace may be affected. Family life, and the other members of the family, are frequently adversely affected by drinking. These problems are not uncommon in present-day Canada. Five per cent of adult Canadians report that their own drinking has caused problems in their friendships or social life in the last 12 months. Three per cent report such problems in their home life or marriage, and two per cent in their work or studies (Eliany, Giesbrecht et al., 1990). From the other side of the relationship, one-quarter of adult Ontarians report that a family member or a friend had a problem with drinking within the last 12 months;

for 10 per cent of all adults, the family member's or friend's drinking had had a "major effect" on them (Room, 1996). The pleasures and pains of drinking are not distributed equally by gender. Men are much more likely than women to drink heavily. But the burden from others' drinking falls disproportionately on women.

We cannot put a price on much of the anguish and heartache from drinking-related problems. But a recent study did add up the costs of Canadian drinking that can be priced (Single, Robson et al., 1996). The total for 1992 was $7.5 billion — over five times the estimated societal cost of illicit drug use. In each Canadian province, the costs of alcohol problems amounted to more than one per cent of the total economy (the GDP). Along with the pleasures from drinking, and the apparent heart disease protection from light drinking, goes a very substantial burden of social and health costs.

A major theme in the story told by Smart and Ogborne is that, throughout Canadian history, drinkers and those around them have struggled with preventing or controlling these costs. At some times and places, these struggles have been primarily in the private spheres of the home and of friendship. But at other times, the "alcohol issue" has entered the public sphere (Gusfield, 1981), on occasion becoming a transcendent political issue. In fact, in the late 19th century Canada was in the forefront of the worldwide temperance movement in terms of popular sentiment in favor of complete prohibition of alcoholic beverages. As Smart and Ogborne note, in a nationwide plebiscite in 1898, a majority of those voting (males only at the time) voted for national Prohibition.

National Prohibition finally did come in Canada, as part of the societal mobilization of the First World War. And, except in Quebec, it was succeeded by provincial Prohibitions of varying lengths. But sooner or later all Canadian provinces retreated from the idea that it was better to ban alcohol outright than to try to limit the harm from drinking.

In place of Prohibition, Canada has moved to a system for reducing the problems from drinking that relies on three main mechanisms. One mechanism is the criminal law and associated regulations, enforcing what might be called situational prohibitions or limits. Notable among these are the drinking-driving laws, forbidding driving with a blood-alcohol level above a fixed limit. While the handling of drinking in the criminal law is a much longer story, drinking-driving laws in their modern form are a post-Second World War phenomenon.

A second major mechanism is the provision of treatment for alcoholism and alcohol problems. As Smart and Ogborne describe, the modern alcohol treatment system is also a relatively recent phenomenon. While Alcoholics Anonymous meetings in Canada date back to the 1940s, the whole modern system of government-supported treatment agencies grew up in the period from the 1950s to the 1980s.

The third major mechanism for reducing the problems from drinking is also the oldest. This is the alcohol control system that exists in each Canadian province and territory. While there are many differences between provinces in the details, there is enough commonality that we may talk of a "Canadian model" of alcohol control. It is a model that in fact was pioneered in Quebec, since Quebec was the first jurisdiction north of the Rio Grande to retreat from prohibition. The model, as it evolved, combined several main elements:

- a relatively high price structure for alcoholic beverages, with distilled spirits taxed more heavily than lower-alcohol beverages;
- a provincial monopoly of the wholesale and retailing of spirits and wine, and often also of beer;
- a strongly regulated licensing system for places for on-premises consumption (restaurants, bars, beer halls, etc.).

The Canadian model of alcohol control proved highly successful. As the United States moved toward repealing Prohibition, a Rockefeller-funded study recommended adoption of the model by U.S. states, and versions of it were in fact adopted by a total of 18 states, many of them along the Canadian border (Room, 1993).

As Smart and Ogborne illustrate from Canadian studies, the modern literature on alcohol control provides evidence that the elements of the Canadian model do in fact help to limit the problems from drinking (Edwards, Anderson et al., 1994). A relatively high price structure holds down the level of drinking, including particularly the drinking of those most at risk of problems. This is also the effect of the restricted opening hours and number of stores which are features of the provincial monopoly systems. A specific licensing system for on-premises sales and consumption makes it possible to hold the restaurant or bar owner accountable for acting so as to limit the problems from drinking.

While it can be argued that the Canadian model of alcohol control has been relatively successful, it is increasingly the victim of its own success. Three generations after the model was adopted, it is easy for the political process to lose sight of the purposes and functions of its various features, and to argue that they have no effect. As in other countries with strong alcohol control systems, in the last 50 years private interests and market forces have successfully pressed to relax many of the provisions of the system. These forces have acted as a "ratchet mechanism" (Mäkelä et al., 1981), pushing the regulatory agenda in only one direction. The details that Smart and Ogborne give in Chapter 8 of the extent of change in alcohol controls in Ontario illustrate that such highly visible changes as the full privatization of retail alcohol sales in Alberta are only part of the story.

By and large, these changes have happened in the teeth of public opinion. Some evidence on the general conservatism of Canadian public opinion about the control system is given in Chapter 8. According to recent Ontario surveys, public support remains high for the main features of the Canadian model of alcohol control. In 1995, 57 per cent of adults in Ontario thought alcohol taxes should remain the same, and 19 per cent that they should be increased. Only 11 per cent felt that beer and liquor store hours should be increased, and only 13 per cent thought that hours of alcoholic beverage service in restaurants, bars, etc. should be increased. Most (73 per cent) thought that alcoholic beverages should *not* be available in corner stores (Anglin, 1995). In an earlier survey, only 13 per cent felt there should be more Brewers Retail stores, while only 12 per cent favored more LCBO liquor stores (Ferris, Templeton and Wong, 1994).

Part of the reason for the discrepancy between public opinion and government actions on alcohol controls may be that alcohol controls find more support among the less powerful. Consistently, strong alcohol controls are supported more by women than by men. While tobacco controls are more favored these days by the highly educated, it is the less educated who more strongly support alcohol controls.

The story this book tells cannot have a neat ending. Drinking is here to stay, and it will continue to bring with it pain as well as pleasure. In their private lives, many Canadians will have to continue to deal with the problems posed by their own or another's drinking. The Canadian story up to this point suggests that alcohol problems will also to some extent remain public problems — that Canadians still expect of their governments such functions as the deterrence of drunk driving and the provision of alcohol treatment services. The future of the great social experiment of the Canadian model of alcohol control is more in question. Public opinion may set some limits to its dismantling. On the other hand, many of its features may disappear in coming years. Ironically, as it is dismantled, researchers will be able yet again to seize the opportunity to measure its effectiveness. But a policy move in the opposite direction, overcoming the ratchet mechanism, would require a new social movement around alcohol issues much stronger than anything on the horizon today.

References

Anglin, L., ed. (1995). *The Ontario Experience of Alcohol and Tobacco: New Focus on Accessibility, Violence and Mandatory Treatment*. Toronto: Addiction Research Foundation, Internal Document No. 122.

Duster, T. (1983). Commentary. In *Alcohol and Disinhibition: Nature and Meaning of the Link*, NIAAA Research Monograph No. 12, ed. Room, R. and Collins, G. Washington, DC: USGPO, DHHS Publication No. (ADM) 83-1246, 326-330.

Edwards, G., Anderson, P., Babor, T. F., Casswell, S., Ferrence, R., Giesbrecht, N., Godfrey, C., Holder, H. D., Lemmens, P., Mäkelä, K., Midanik, L. T., Norström, T., Österberg, E., Romelsjö, A., Room, R., Simpura, J. and Skog, O. J. (1994). *Alcohol Policy and the Public Good*. Oxford: Oxford University Press.

Eliany, M., Giesbrecht, N., Nelson, M., Wellman, B. and Wortley, S. eds. (1990). *National Alcohol and Drugs Survey (1989): Highlights Report*. Ottawa: Health and Welfare Canada.

Ferris, J., Room, R. and Giesbrecht, N. (1993). Public health interests in trade agreements on alcoholic beverages in North America. *Alcohol, Health and Research World* 17: 235-241.

Ferris, J., Templeton, L. and Wong, S. (1994). *Alcohol, Tobacco and Marijuana: Use, Norms, Problems and Policy Attitudes among Ontario Adults*. Toronto: Addiction Research Foundation, Internal Document No. 118.

Gusfield, J. R. (1981). *The Culture of Public Problems: Drinking-Driving and the Symbolic Order*. Chicago & London: University of Chicago Press.

Mäkelä, K., Room, R., Single, E., Sulkunen, P. and Walsh, B. with 13 others (1981). *Alcohol, Society, and the State: 1. A Comparative Study of Alcohol Control*. Toronto: Addiction Research Foundation, 1981.

Room, R. (1993). The evolution of alcohol monopolies and their relevance for public health. *Contemporary Drug Problems* 20: 169-187.

Room, R. (1996). Patterns of family responses to alcohol and tobacco problems. *Drug and Alcohol Review* 15: 171-181.

Single, E., Robson, L., Xie, X. and Rehm, J. (1996). *The Costs of Substance Abuse in Canada*. Ottawa: Canadian Centre on Substance Abuse.

Preface to the Second Edition

A decade has passed since we started to work on the first edition of this book and much of interest has happened in the interim. This second edition of *Northern Spirits* aims to describe the main changes that have taken place with respect to drinking patterns, problems and policies and to draw attention to recent, relevant research. The second edition also reflects comments and suggestions from readers of the first edition and changes in our own thinking about drinking and drinking problems in Canada. A chapter on the treatment of alcohol problems in Canada has been added and several other chapters in the first edition have been extensively revised.

We are grateful to our colleagues at the Addiction Research Foundation for their support and assistance in the preparation of the revised manuscript and especially to Jürgen Rehm and Robin Room for their helpful and insightful comments on the first draft. Barb Keogan and Susan Steinback worked wonders at their word processors and Gail Schmidt relished her task as a proof reader of early drafts. Special thanks also to Myles Magner for editing the final draft of this edition and to Nancy Leung, Paula Poschun and Bob Tarjan for its design and production.

Preface to First Edition

Alcohol has created problems in Canada from the time of the earliest settlers in New France. There has been no period without problems of drunkenness, alcohol-related health problems such as liver cirrhosis, and alcohol-related violence. Even during the short-lived period of Prohibition such problems continued, albeit at a diminished level.

Concerns about alcohol problems have changed drastically since the early days of Canada. Then the main concern was drunkenness and violence. Now the greatest concerns are alcohol-related accidents and youthful drinking. Our special problems with native populations appeared early and have never been satisfactorily solved. The aim of this book is to describe the alcohol problem in Canada and the methods of prevention that have been tried.

Much of what has been done about alcohol in Canada reflects the puritan background of the country's early days. However, a study of alcohol problems puts us in touch with the major themes in Canadian life — the conflicts between English and French Canada, the federal-provincial conflicts for jurisdiction, the regionalism and provincialism of social life, and the tension between government control over personal matters (such as drinking) and individual freedoms. The study of alcohol problems provides us with a mirror in which we can see ourselves more clearly.

How Canada has assessed and reacted to alcohol problems reflects the culture, the pathways, and the mores in a way that is interesting and worthy of study. The approach taken has been different from the "law and order" approach taken with illicit drugs. A social control model with carefully chosen democratic controls and constant reference to public opinion has been used in instituting the alcohol control mechanisms. In addition, Canada has become a world leader in alcohol research. Chiefly through the Addiction Research Foundation of Ontario, the British Columbia Alcoholism Foundation, Health and Welfare Canada, and a few university researchers, alcohol problems have been described and numerous solutions proposed. South of the border, at least, there is a "Canadian approach" to the prevention of alcohol problems, which requires that per capita alcohol consumption be reduced through government controls on availability.

Despite the contributions of Canadians to alcohol research, there is no one source of information on Canadian alcohol problems. There is not a single book on the topic, although recently Health and Welfare

Canada published a long report (*Alcohol in Canada: A National Perspective*). The Le Dain Commission, a decade ago, made a desultory review of the situation, but little has been done since to review and integrate various findings. Much of the early material on the temperance movement is buried in libraries or rare historical documents, and some of it is unclassified. Several theses have examined the temperance movement in Canada, but they are difficult to obtain except on microfilm. Books on Prohibition such as Spence's *Prohibition in Canada* and Hallowell's *Prohibition in Ontario* are long out of print. Much of the modern material on alcohol problems is scattered, difficult to find, and frequently unanalysed in detailed statistical reports. The aim of this book is to describe the history and present status of drinking and drinking problems in Canada, and to consider ways in which drinking problems can be prevented.

The authors wish to acknowledge the help of the following people and institutions: Deborah J. Lindholm, Wendy Rush, Barbara Keogan, and Michelle Hoo for typing various drafts of the manuscript; Mario Faveri and John La Rocque for allowing Alan Ogborne time to work on this book; Catherine Cragg for editing the manuscript; Lia Hatashita for co-ordinating production; Donald Murray for designing the book; Robert E. Popham for commenting on early drafts of the historical sections; the Addiction Research Foundation library for making a special effort to find materials; the Women's Christian Temperance Union, Alcohol and Drug Concerns, Sons of Temperance, and Victoria University Library for materials on the Temperance Movement; Irving Rootman for materials on prevention programs in Canada; the Public Archives of Canada for the photographs used; and Jane Ogborne and Lillian Smart for their patience.

Chapter 1

Pioneer Drinking and Drunkenness

Before Europeans came to North America, Canada was bone dry and its inhabitants were totally abstinent. In pre-Columbian times, neither native Indians nor Inuit had discovered fermentation, although it was well known in Mexico and other parts of Latin America. Just why the more northern indigenous peoples did not discover that fermenting fruits and vegetables could produce an intoxicating beverage is a matter for speculation. It has been suggested that the nomadic pattern of aboriginal life with its dependence on hunting and slash-and-burn agriculture made for an unsettled people, who were never in one place long enough to develop techniques for producing alcohol.

Furs for Drunkenness

The first Europeans in Canada were French explorers, adventurers and traders. They were shortly followed by missionaries and a few settlers who lived mainly in the St. Lawrence area of Quebec. *Coureurs de bois* made long trips by canoe in trading with Indians for furs. In addition, Indians came to French forts seeking to trade furs for blankets, guns, beads and other European goods. Alcohol played a significant role in the North American fur trade. Brandy was a major commodity traded to Indians in exchange for fur, and in many areas of the country no trade of any type would be conducted without it (Heath, 1982; Ray, 1974; Rich, 1967).

The introduction of alcohol had an early and dramatic impact on Canada's aboriginal people. According to contemporary accounts by Europeans, from a very early time some Indians showed a great interest in alcohol and a tendency to get drunk on each drinking occasion — a pattern that often resulted in violence. In a typically lurid description, one 17th-century Quebec observer wrote:

> Blood flows at the feast. The women get drunk like the men and resemble veritable furies. Nothing more horrible can be imagined than a wigwam of braves, awakening the day after the carousal to

> find themselves disfigured, crushed, surrounded sometimes by the dead bodies of their relatives or of their friends (Rochemontaix, 1919).

There is evidence, however, that some Indians did not like alcohol at first. Howay states that the first whisky was brought to the northwest coast by traders and that Indians spat it out as an unfamiliar and unpleasant drink (Howay, 1942). Traders had to work to develop the trade in whisky and teach Indians to drink and become drunk.

Reports by early missionaries such as those contained in the *Jesuit Relations,* written in the 17th century, include hundreds of references to drunkenness and the impact of alcohol on aboriginal communities (Howay, 1942). Observations that "every night is filled with clamours, brawls and fatal accidents, which the intoxicated cause in the cabins," and that "it [drunkenness] is so common here and causes such disorders, that it sometimes seems as if all the people of the village had become insane so great is the licence they allow themselves when they are under the influence of liquor," are typical (Dailey, 1968). As well as violence, increases in family disunity, seduction of innocents, and starvation in native communities were attributed to alcohol.

Very early in the fur trade, Europeans observed certain patterns of drinking among native people. Some Indians had developed the habit of consuming all available food at one sitting, and "eat-it-all feasts" were common (Dailey, 1968). When alcohol was introduced, the same approach was taken in "drink-it-all" sessions, sometimes lasting several days or weeks (Carpenter, 1959). The introduction of alcohol also appears to have had an impact on native spiritual practices. Among the Iroquois, dream or vision quests were mystical experiences in which a dream-like state was induced through seclusion, fasting and meditation. After the introduction of alcohol, however, there is evidence that some Iroquois adopted alcohol intoxication as a way of achieving such a spiritual state. Accounts from the time also describe situations in which several Iroquois would combine their limited supplies so that at least some of their number could drink enough to experience a dream quest (Carpenter, 1959). However, others have suggested (MacAndrew and Edgerton, 1969) that drinking among native people was not, in fact, much different from that of the early European traders. Nonetheless, drinking was likely greater in some native communities than it was among early settlers and farmers, and differed considerably from the dominant drinking styles in Europe.

French authorities made several efforts to control drinking by Indians. Jesuits frequently spoke out against such drinking, and in

1657 Louis XIV issued an edict prohibiting the sale of alcohol to native people. Although fines and public flogging were the penalties, the practice continued, in part because much trade took place so far from civil authorities. In 1660 Bishop Laval issued a sentence of excommunication on those trading alcohol to native people. The sentence was extremely unpopular, however, and was later rescinded because some French authorities believed that the trade in brandy was essential to retain the Indians' allegiance against the English (Spence, 1919).

Dutch and English traders and the "gentlemen adventurers trading into Hudson's Bay" (the Hudson's Bay Company) cared little about excommunication from the Catholic church. Rum brought by English traders from the West Indies helped to secure them a larger part of the trade in furs. By 1800 virtually all of the fur trade in eastern Canada had fallen into English hands because rum was made so readily available (Rich, 1967). Whenever efforts were made to control native people's access to alcohol, the traders pointed out that some other group of traders would gladly fill the demand. For example, in 1775 Samuel Hearne, in charge of the Hudson's Bay post at Cumberland House, claimed that:

> For though we undersell the Canadians (Montreal pedlars) by far in some articles while our goods last, yet when Brandy is out the Indians leave off trading. Last Spring . . . the Indians would not trade their furs because I had no brandy (Rich, 1967).

Similar statements were made by officials of the Buffalo Wool Company, trading in Manitoba in competition with French, Dutch, and American traders (Garland and Talman, 1931). Records compiled by Ray (1974) show that at some places in the Northwest alcohol was a small part of the trade with Indians but at others it was the main item Indians sought. For example, one 18th-century account noted "the only thing which was necessary to the conduct of the trade there [Brandon House] was a good supply of rum." From time to time various trading groups, including the Hudson's Bay Company, banned the alcohol trade with native people. Such bans were rarely enforced until 1821 when the Hudson's Bay Company and the North West Company combined. After the union, far less alcohol was given or traded to Indians, in keeping with company policy (Oliver, 1923).

Brandy and rum were frequently given to the native people with no exchange necessary. In many trading posts Indians would be given large amounts of alcohol as a "gift" before trading began. Traders found that far better deals could be struck once their trading partners were intoxicated. Presenting alcohol as gifts also helped traders evade the various edicts that forbade "trade" in alcohol. Since large trading companies often had a monopoly, Indians could not go elsewhere with their furs.

For many years, governments included supplies of alcoholic beverages in their gifts to Indians on "treaty days." In some areas this practice was stopped by the late 1820s at the request of band chiefs (Garland and Talman, 1931). However, the annual presents were often bartered for whisky. In the 1840s an edict was passed in Upper Canada forbidding the sale, gift, exchange or barter of alcoholic beverages to Indians. Nonetheless, almost all early settlers' diaries contain many references to intoxication and violence among native people.

The role of alcohol in the fur trade — and the resulting harm to native communities — was only one aspect of its impact on pre-Confederation Canada. Alcohol also had an important place in the developing social and economic life of Canada's expanding European settlements.

Early Alcohol Controls

After the British conquest of Canada in 1759, liquor was looked upon mainly from the point of view of profit. In Nova Scotia, Governor Belcher said, "authorities are very desirous of a very large consumption [of alcohol] to increase the income of the province" (Spence, 1919). By 1774, all alcohol control regulations imposed by the French had been repealed and replaced with a licensing system. Licences to sell alcoholic beverages were readily given throughout the country in the early 1800s, and many illegal sources also existed. Licences were originally obtained only from the governor of the colony, but by 1794 they were given by local magistrates, who were to ensure that tavern keepers kept a "decent and orderly" house. Most early regulations dealt with licence fees and their collection and how the money could be spent, not with drunkenness or other controls on consumption (Popham, 1976). Accounts from early settlers agree that few regulations were enforced, and obtaining a licence for an inn or tavern was an easy matter.

One interesting early experiment in alcohol controls did occur as "the first Canadian war-time prohibition measure" (Riddell, 1920). During the war of 1812, many farmers were called into the militia and much less wheat was planted. At that time wheat was made into flour or whisky and distillers offered higher prices for grain than millers. In order to preserve grain for flour, the government of Upper Canada prohibited the distillation of spirits from grain for about a year. Complaints were made that soldiers could not fight without whisky, and special provisions were made for them. The legislation lapsed in 1814. In its limited nature it foretold what later prohibitions in Canada would be like — half-serious attempts at control for short periods.

Drinking in Early Settlements (prior to 1840)

In the years prior to 1840, the population was small, spread over vast areas, and predominantly rural. There were few cities of any size (except for Toronto, Montreal, and Halifax), and none west of Ontario (known before 1840 as Upper Canada and after 1840 as Canada West). The settlers were virtually all new to the country (fewer than half had been there 10 years) and were struggling with the innumerable problems of starting subsistence farming in a hostile environment. Problems of clearing the land, planting crops, enduring the winter without proper clothing, and preventing crop failures, fires or other calamities wholly occupied most settlers, leaving little time for what we call recreation. However, considerable evidence exists to show that the early settlers imbibed, and frequently to excess.*

Our knowledge of exactly how much was drunk in the early days is incomplete. Of course there were no drinking surveys, nor even a good census until 1851. Details about how much beer, wine, and spirits were made or imported are also fragmentary. However, we do have extensive information about inns and taverns. Best of all, we have the firsthand accounts of settlers themselves. As well, travellers such as Joseph Howe, William Hickey, Patrick Shirreff, John Howison,

---

* Some of the best accounts are the following: E. C. Guillet, 1979; A. Jameson, 1943; S. Strickland, 1970; S. Moodie, 1959; S. Moodie, 1980. Much of our information about pioneer drinking comes from the writings of the Strickland family; Susanna (Mrs. Moodie), Catherine (Mrs. Trail), and Samuel all wrote books about their experiences in Canada. All had extensive experience in Upper Canada and all paid some attention to local drinking habits. They came from a well-to-do family but faced an enormous drop in living standards in Canada, living for years in primitive shacks in underdeveloped areas and seeing pioneer life at its worst. Samuel and Susanna probably best described pioneer drinking habits. Susanna was, in addition to a backwoods farm wife, a romantic novelist. She was a staid, serious woman much concerned with impropriety, and her accounts of excessive drinking among pioneers might be overdrawn. Samuel had a wider experience in Upper Canada and appears more tolerant of human weakness. His accounts of drinking among pioneers are probably more balanced.

James Pickering and others wrote travelogues that were intended as guides for potential emigrants in Britain.* Generally, such books described every aspect of life, including farming, social life, recreation, and mores and customs. It is from these that we gather most of our information about early drinking and its problems.

The first brewery in Canada is said to have been established in 1668 (Brown, Devan and Wallace, 1982). However, most early settlers drank whisky made from wheat, rum imported from the Caribbean, or brandy from France. A "white" whisky was the most preferred beverage in the early 1800s in Upper Canada (Riddell, 1920).

Most settlers' accounts and immigrant guides point to extensive heavy alcohol consumption, especially by men. For example, Susanna Moodie avowed that "this frightful vice of drinking prevails throughout the colony to an alarming extent. Professional gentlemen are not ashamed of being seen issuing from the barroom of a tavern early in the morning or of being caught reeling home from the same sink of iniquity late at night." She gave examples of two physicians "who entered upon a contract to drink until they both died" and then succeeded in their project. She even gave an account of an itinerant alcoholic, probably the first account of that type in Canada.

Men of prestige were often heavy imbibers. For example, Colonel Thomas Talbot, the founder of the Talbot Settlement and a magistrate, was reported to be a drunkard and the "worse of liquor by 3:00 p.m. when he was rude and insulting" (Garland and Talman, 1931). Talbot condemned the temperance societies as "damned cold water drinking societies" and became an enemy of the temperance movement (Barron, 1976). Magistrates holding court sessions in taverns (often the only available building) frequently drank alcohol during the proceedings. An early report describes an expense account from the Dundas Assizes in 1790 that includes "liker for the gentlemen of the grand-jury and for decanters broken" (Garland and Talman, 1931).

Drinking was not frowned upon among the clergy and churchgoers until the temperance movement. The Presbyterians in 1792 gave a gift of two hogsheads of Spanish wine to the Recollect Order for use of the building (Elgee, 1964). No denominations required abstinence before 1830, even Methodists, Baptists, and Presbyterians, the eventual supporters of the temperance movement. Up until 1847, church newspapers for the Anglicans and Presbyterians took advertisements for alcoholic beverages. Bishop Strachan, the Anglican Bishop of

* See, for example: W. Hickey, 1834; J. Howe, 1973; J. Howison, 1821; J. Pickering, 1831.

Toronto, was known to always travel with crystal glasses and a decanter of brandy in his carriage. He admonished servants for overfilling glasses of wine at official functions and took great care not to have any spilled. He too became a foe of the temperance movement.

Heavy drinking was often done at festivals, special events, elections, and pioneer "bees." For example, W.H. Merritt, a resident of Welland area, complained that as a result of liquor, "at every bee, every militia training, even at our dances swearing, cursing, quarrelling, fighting, biting and even gouging, was a common occurrence" (Merritt, 1875). He also described an election that "occurred on the 1st of July 1836 and . . . for riot and drunkenness exceeded everything I have ever seen before and challenges the world in this peculiar line." With regard to militia parades, the only disagreement was "how much drill was necessary before the drinking and horse racing began." Drunkenness at such events was a common occurrence, according to both Pickering and Proudfoot (undated), and greatly desired by the participants. Heavy drinking at marriages and deaths is well described by Howison in his *Sketches of Upper Canada (1821).* The custom was to offer liquor to mourners and friends at funerals, and sometimes even to the corpse. Marriages also required music and dancing and often a large amount of whisky.

Drinking and Pioneer Bees

Pioneer bees deserve special attention as occasions for drinking. These were events that brought together members of a particular town or area to complete some large work project. The most common bees included logging, barn raising, shanty raising, land clearing, lime burning (for mortar) and mowing — all work that involved primarily men. There were baking, quilting, spinning and apple-paring bees for women, and they often were the occasions for young people to get to know each other. Bees had the purpose of accomplishing the hard work that could not be done by a single family, such as raising a barn or building a house. However, they also had important social functions. They brought people together over serious work and fostered a feeling of co-operation in a hostile environment. Not unexpectedly, they were occasions for socializing, hilarity and heavy drinking. During the months of May to November, most settled areas had bees of some sort about every week.

Most large bees were expected to end in a dance for any who were still sober. The best ones provided important opportunities for courting for young people and recreation for older people. At all bees there was abundant food and drink. Some bees such as apple-paring bees and pumpkin bees (baking of pies) were known as young people's events; much fun and little work were expected.

Bees played a central role in making pioneer life possible. Virtually all early barns, roads and houses were built as a result of them. They were particularly important in new communities when few settlers had established houses or barns. Of course, they were not popular with everyone. Susanna Moodie said that "people in the woods have a craze for giving and going to bees . . . plenty of strong drink and excitement making the chief attraction of the bees." The Moodie family had particularly bad experiences with logging bees, which, along with barn-raising bees, seem to have attracted the heaviest drinkers.

Anyone putting on a logging, mowing or raising bee was expected to supply abundant food and liquor. In fact, if liquor was not supplied, neighbors would frequently not come. According to Guillet's *Pioneer Days in Upper Canada* (1979), it was the custom to provide a gallon of whisky per pair of oxen — that is, per attending family — at each bee, or else an unlimited supply. Samuel Strickland described a mowing and cradling (scathing) bee at Darlington, Ontario, where the "grog boss passed out water and whisky continuously." Every bee had a "grog boss" whose job was to make sure that there was sufficient whisky supplied to get the work done. Meals were expected to be heavy and to be accompanied by plenty of beer and whisky. Strickland's wise advice to the new emigrant was that "no liquor should be served out to the swarm of working bees until the raising is over as many serious accidents having occurred for want of this precaution." Extra precautions seem to have been taken with raising bees, but a large number of deaths from drunken falls were reported. More than 80 gallons of whisky were consumed at one bee described by Strickland.

Not all bees resulted in the same drunken and riotous behavior. In his *Hints on Emigration to Upper Canada*, William Hickey said that food was not necessary for bees if the settler was poor — "only some whisky and the evening frolic are sufficient inducements for the attendance of your neighbors." Bees to raise the homes of ministers of the church did not require food or drink to be provided, although settlers sometimes brought their own. Quilting, paring and spinning bees, which were attended by women and young people, apparently did not involve much whisky or drunkenness. One of the first roles undertaken by the temperance societies was to sponsor bees at which there was no drinking. Temperance society members would go only to bees of that type, and the practice of having drunken bees was greatly reduced by the 1860s.

Early Inns and Taverns and the Promotion of Drinking

Not all the drinking by early settlers occurred at bees and festivals. Much of it was done at the numerous inns and taverns that dotted the landscape. Prior to the coming of the railroads, virtually all land travel was by horse or by horse and cart. Because horses could not travel far (25 miles per day was a good trip), wayside inns were needed every few miles to provide food and lodging, food for horses, or fresh horses. Virtually all inns also had taverns, and they promoted heavy drinking both by their numbers and their ways of doing business. Taverns also provided many kinds of social contacts and places for meetings of all sorts.

The number of taverns prior to the temperance movement was very large. For example, in the 1840s the traveller from Toronto to Barrie would pass 66 licensed taverns, plus a number of unlicensed ones (Guillet, 1954). This would be more than one per mile. The first official census of Upper Canada in 1851 counted 1,990 taverns, or one for every 478 people. In 1995, in Metro Toronto, a city well supplied with drinking establishments serving both residents and visitors, there were 3,796 licensed bars and restaurants, or one for every 600 residents. Anna Jameson stated that in the 1830s London, Ontario, had seven taverns for 1,300 people, plus a number of grocery stores that were "grog shops" or drinking houses. Goderich, Ontario, in 1833 had only 40 people but three taverns (Garland and Talman, 1931). Tavern licences were granted readily by magistrates, and all that seemed to be needed was the desire to open one.

In addition to taverns, many stores sold alcoholic beverages and allowed drinking by customers. Others gave it away, as in the 1820s whisky was very cheap (half a penny per gallon). According to a clergyman named Weir, most stores in Ontario kept a pail of whisky and a tin at the back door where customers could help themselves at no expense (Guillet, 1979).

Heavy drinking seemed to be the rule in pioneer taverns, as it was described by numerous settlers. One emigrant's advice book by an anonymous ex-settler warned that "immigrants will find every inn and tavern and beer shop filled at all hours with drunken, brawling fellows and the quantity of ardent spirits consumed will truly astonish you." Patrick Shirreff in his *Tour Through North America* in 1835 said that "the barrooms of the hotels were filled with swearing, tipsy people" (Garland and Talman, 1931). Two remarks stood out in the letters of Sussex emigrants who came to Canada in 1831: the liquor was cheap and enormous amounts were drunk. Pickering noted that the

Upper Canada inns of the 1820s charged only for board and lodging and not for whisky. Alcohol consumption must have been substantial, although detailed amounts are difficult to determine.

The Importance of Early Inns and Taverns

Much of the social and cultural life of early settlers centred around taverns. They were not merely drinking places. In new settlements both the school and church were typically built after the tavern, which was the first building put up. Most were open seven days a week for travellers, who often brought outside news to a population that rarely saw even a newspaper. The tavern barroom provided a warm fire on a cold night and the only place for discussion and debate about events of the day. As Anna Jameson said, "taverns and low drinking places" were the only place of assembly or amusement in the 1830s and "taverns consequently abounded everywhere."

Meetings of all sorts took place in taverns. The larger ones had ballrooms that accommodated weddings, government balls, local dances, militia meetings, and other banquets. Visiting circuses performed at taverns. Wages of agricultural and other workers were paid in taverns. A host of social, religious and political groups, such as St. George's Societies, agricultural societies and political parties, met there (Barron, 1976). Only the temperance organizations refused to meet in them.

In many areas taverns became the first quasi-official government buildings. Electioneering and polling both were held there. Larger taverns often became the local courts and held the offices of magistrates. Tavern keepers were well-respected members of society and had both high social status and popularity. In pioneer times it was no disgrace to marry a tavern keeper, and the bride could be sure of a comfortable life.

Bishop Strachan often preached in taverns on his tours of rural areas, as did several Presbyterian ministers. Since they were built before churches, sometimes church services and Sunday schools were held in taverns and inns (Barron, 1976).

Because taverns were known as places for debate and dissent, new political movements often had their beginnings in the barroom. It is well known that the 1837 Upper Canada Rebellion led by William Lyon Mackenzie was launched from his brother-in-law's tavern.

Since the tavern occupied such an important place in social life, it is little wonder that tavern drinking became so important. Almost any meeting or festivity was likely to be located in a tavern, and hence it would seem natural to have alcoholic beverages for the occasion. As we will see later, part of the temperance movement became more "anti-tavern" than "anti-drinking." Not until the temperance movement began did the pioneer tavern have any real opposition, nor was there any alternative site for the many social events it accommodated.*

How Much Did People Drink in Pioneer Times?

There is no doubt that a large amount of heavy drinking was done in pioneer times. However, we have great difficulty determining just how much, and how it compares to today's drinking. Although numerous reports of heavy drinking were made by some writers, others made no report at all. For example, Alexander Ross in his account of the Red River settlement does not mention drinking or drunkenness, although he describes every other aspect of life (Ross, 1972). Joseph Howe made several trips over Nova Scotia in 1828-1831 and described some 16 taverns and inns, but makes no mention of drinking or drunkenness, although he was a very fastidious person. However, heavy drinking did seem to characterize many settlements in the Maritimes and Upper and Lower Canada. Drunkenness was prominently mentioned in most settlers' accounts.

Because so much of pioneer drinking seems to have involved large amounts on special occasions, the total amount consumed may be overestimated. There is a tendency to think perhaps that pioneers were always drunk. However, several factors would work to reduce the overall amount of pioneer drinking. One would be the low disposable income; money for non-essentials was rare. Most important, the difficult lives of most settlers and the large amount of work they had to do would not allow much drunkenness except on special occasions.

Another consideration is that nearly all early records of drinking involve men. Accounts of female drinking or drunkenness from before 1850 are almost non-existent. All of the cases of "alcoholism" among settlers described by Moodie, Jameson and others were men. The pattern seemed to be one of periodic heavy recreational drinking. There is a strong suggestion of "time-out" drunkenness, the seeking of release from the stressful lives that faced early settlers.

Most of the alcohol drunk in those days was whisky, brandy or rum. According to Guillet, in the early 1800s farmers did not make beer or wine but had some grain made into whisky at the local mill.

---

* Readers who wish to see an early Ontario tavern in its original form should visit Montgomery's Inn in Etobicoke. The barroom is virtually the only one remaining in Ontario from the 1840s. Incidentally, this is not the Montgomery's Tavern involved in the Upper Canada Rebellion.

There were no local, and few imported, wines, and beer was not drunk much until the 1840s. Whisky was typically consumed periodically and in large amounts, rather than a few drinks each day. Drinking with meals is rarely mentioned in settlers' accounts unless the reference is to special events, such as bees, banquets or militia dinners. Daily heavy drinking seems to have been rare.

We have only an incomplete picture of per capita alcohol consumption in times prior to 1850. Proudfoot expressed the opinion that there were few men who drank less than four glasses of whisky a day, but that amount is almost certainly too high for the population as a whole. Also, the rate of abstention is difficult to judge. Some believe that everyone drank, but not everyone drinks even today. Also, evidence from many modern surveys has shown that abstaining from alcohol is much more common among rural people; given that in 1851 85 per cent of the population lived in rural areas, a fair proportion of abstainers could be expected. Other abstaining groups would include children who, because of the large families and shorter average life span, would have made up a large part — perhaps as much as 65 per cent — of the population.

Some idea of per capita consumption can be obtained from the census of 1851, the first in Canada West. It listed 49 breweries, which made 750,000 gallons of beer, and distilleries, which made 2,921,700 gallons of liquor, for a population of 952,000. Barron (1976) assumes in his study of temperance in Upper Canada that some 400,000 gallons of wine and spirits were imported and that, altogether, per capita consumption of wine and spirits was 3.49 gallons per person. This is higher than his figure for Ontario in 1976, which was 3.12 gallons per capita. In addition, Barron has argued that spirits contained twice as much alcohol in the 1850s as they do today. While this level is unlikely, as beverages with 80 to 90 per cent alcohol would not be very palatable, alcohol content was not government-regulated and was probably somewhat higher on average than today. Since the proportion of children (non-consumers) was higher in the 1850s, and abstainers were probably more common because of the temperance movement, per capita consumption of absolute alcohol by those who did drink was certainly higher, perhaps by 25 or 30 per cent, than today.

Such an estimate is consistent with studies of drinking in other 19th-century colonial societies. Powell (1988) made similar estimates for drinking in New South Wales, Australia in the 1820s. He showed that per capita consumption, corrected for the proportion of children and females, was about 19 litres of absolute alcohol per person compared to 15.6 in 1977-78.

Drunkenness and the Male Character of the Early Frontier

We have been describing drinking in Upper Canada, chiefly among settlers and pioneers. Upper and Lower Canada and the Maritime provinces had a predominantly rural life in the early 1800s, characterized by small settlements of farmers and their families who tilled the land. This stable, family-oriented way of life contributed to the early development of the temperance movement and other types of moral reform.

However, much of Canada was outside this settled pattern in the 1800s and closer to the wide-open frontiers later romanticized in Hollywood movies. Large parts of the country had virtually no farmers or settlers. The Northwest Territories, British Columbia and large parts of Alberta were not settled. In the 1800s many of the European "communities" in such places were peopled mainly by men who worked dangerous occupations such as lumbering, mining, and heavy construction of roads and railways. These isolated communities often had no family life and included few women who were not prostitutes. The work often meant feverish activity in the short summers and long, dull winters with little to do but go to the saloon.

Lumbermen acquired a particularly bad reputation for excessive drinking and violence (Cross, 1960). McGregor (1833) described a typical camp where each lumberman had a drink of raw whisky before breakfast. Eventually, employers excluded alcohol from all lumbering shanty towns, and alcohol problems were greatly reduced by about 1840. However, lumbermen worked only eight months of the year, and when the log drive down river began, "in the matter of scenting out and appropriating whisky the lumberman was as keen as a weasel." Saloons every few miles along the Ottawa River sold lumbermen bad whisky, and many farms were terrorized on the way. When they arrived in Montreal, they squandered all their money on "the fiddle, the female and the firewater." The Reverend A.W. Rose, (1849), who followed the log drive, said that the raftsman "sets up for a gentleman . . . till his cash is gone, his health perhaps shaken; he parts with his gay apparel, if it has not been already destroyed in some drunken row, shoulders his axe, and sets off again to the wilderness."

Excessive use of alcohol was also a feature of most early mining and gold-rush towns. Many men in such towns expected to be there for only a short period and looked on it as a grand spree. This transient, unstable population made it difficult to impose social controls. One observer said that "it would almost take a line of packet ships running between here and San Francisco to supply this Island [Vancouver] with grog" (Hiebert, 1969).

Similar social conditions encouraged heavy drinking when the Canadian Pacific Railway was being built in the 1880s. Work camps were usually temporary, set up far from large centres and composed entirely of men. The work was long and dangerous, but days off were livened up by visits to local saloons. Herbert Gowen (1899) described the town of Yale during the building of the CPR: "Drunkenness and disorder filled the place day and night. Fire kindled by lights held in hands unsteady with drink were of almost daily occurrence, the jail was overflowing and the justices weary. Tattered, dirt-bespotted drunkards rolled about the streets, wallowing in the mud, cursing and fighting and driving all respectable people into the recesses of their homes." Of course, bender drinking on days off affected work, and eventually a "dry" area was created around the construction. Originally it was a 20-mile belt, but later it was broadened to 40 miles. A similar dry belt was created for the Grand Trunk Pacific line, but bootleggers paid little attention to it.

Summary

Before Canada was settled by Europeans there was no drinking, but soon afterward drinking became extensive. Heavy drinking was common among Europeans and Indians involved in the early fur trade. Although many efforts were made to control it, excessive drinking reached phenomenal levels among some native people as a result of alcohol's prominent role in the fur trade. Early settlers, especially males, were also heavy drinkers. Most settlers' accounts emphasize the number of taverns and inns and the drunkenness of the population. Drinking seemed to pervade most festivals, special events and pioneer bees. Early inns became the focus of social life. Per capita alcohol consumption was probably 25 to 30 per cent higher than it is today. Most people drank whisky, probably stronger whisky than we have now. Of course, the male character of frontier life in the West and North also contributed to heavy drinking by railwaymen, boatmen, lumbermen, and miners.

Chapter 2

The Temperance Movement: A Drastic Cure for Pioneer Drunkenness

Few social forces in the history of Canada can match the temperance movement for longevity and success. The movement began in the 1820s, and a century later had achieved its goals of converting a large part of the population to abstinence and getting Prohibition introduced in all areas of the country. For a grassroots movement with no government funding and much opposition, its achievements were remarkable. We tend now to think of temperance as outmoded, a dusty remnant of a Victorian past, or as a long, tedious effort to limit drinking and good times. However, the temperance movement had both good and varied features, especially in the early days. It provided a wide variety of social opportunities for its members, many of whom were concerned with general social problems as well as with drinking.

Few evaluations of the temperance movement in Canada have been made. There are several historical papers and theses on the topic but no overall history except by temperance workers themselves such as Spence (1919). However, a recent book by Warsh (1993) contains several valuable papers on the temperance movement in Quebec and the Maritimes and on Prohibition in British Columbia. Also, Gray's book (1982) gives information on drinking problems and Prohibition in Western Canada.

The Canadian temperance movements were roughly modelled on those in Britain and the United States. However, the Canadian movement was more successful than those in Britain and the United States: the former never achieved Prohibition while the latter did so considerably later than Canada. In this chapter we examine the history of the temperance movement and its messages, the reactions to the movement, its benefits, and its connections with other social reforms. The decline of the movement is also examined.

The Origins of the Canadian Temperance Movement

The date of the first Canadian temperance meeting is somewhat in doubt. Various authorities list it as having been at Beaver River, Nova Scotia on April 25, 1828 (Spence, 1919; Garland and Talman, 1931). However, claims have been made that a temperance meeting was held in Montreal in June or July of 1827. Whatever the exact date, it is clear that the movement spread rapidly throughout the Maritimes and Lower and Upper Canada. By 1832 there were 100 societies in Upper Canada with 10,000 members, and they were served by 12 temperance newspapers. By 1842 there were 166 temperance societies with 60,000 members; more than one person in 10 was a member.

Much of the impetus for the temperance movement in Canada came from the United States rather than Britain. The first society in North America was founded in Virginia in 1807, and other societies were soon organized in New Jersey, Massachusetts, and New York State. Many of the early itinerant speakers on temperance were American. At once, opponents of the temperance movement, such as Colonel Talbot, damned the societies as a product of "Yankeeism" (Barron, 1976). Canadian churches such as the Presbyterian, Methodist, Episcopal, and Baptist, which often got their clergy from the United States, typically supported temperance. Those getting their clergy from England, such as the Anglicans, typically did not support it. It has been pointed out by Hiebert that the temperance movement came latest to British Columbia and had little influence there because that province traditionally had the largest proportion of Anglicans (Hiebert, 1969). The Catholic Church also took little interest in temperance in the 1800s.

The close association of the temperance movement with Protestant fundamentalist and evangelical churches began early and continued throughout its history. In Upper Canada, most of the early temperance societies were organized by Methodists, who were the most numerous Protestant religious group in the 1830s and had close associations with the American Methodist churches. In the Maritimes, Baptist churches organized most temperance societies. Later, Presbyterians became active in the movement, and numerous churches were formed where the minister and the whole congregation took the pledge against drinking (Elgee, 1964). Whole congregations of Methodists also took the pledge, and it came to be understood that the temperance movement and Methodism were synonymous in Ontario. In the 1840s, one could not be a good Methodist without joining a temperance society. When the temperance movement went west in the 1870s, it was the Methodists who took it there.

It was different for the Anglican Church. It included people who were both Christians and "gentlemen," and gentlemen were not teetotallers (Elgee, 1964). Very few Anglican ministers took any role in the temperance movement, and virtually no societies were officially supported by the Anglican Church. In the pioneer era, many Anglicans thought that "religion had too much fanaticism" and needed a quieter, more tolerant and traditional stand. Many brewers, distillers, and holders of licences for hotels were Anglicans, and their opposition to temperance was understandable. It is true, of course, that the Anglicans and their church were both against drunkenness and in favor of only moderate drinking. However, with the temperance movement, fundamentalist churches, especially the Methodists and Baptists, established "ownership" of drinking problems in Canada. They came to define what those problems were, who were the chief actors, and what could be done about them. It was to be almost a hundred years before governments, the medical profession, and other professionals began to challenge this ownership.

Early Temperance Societies

The first temperance societies were local and small in size, usually being limited to a single town or district. One of the earliest societies in Upper Canada was the temperance Society of Ancaster and Beverly, founded by Egerton Ryerson and John Rolph. Its purpose was to "restrain the use of ardent spirits to cases in which the use of them may be recommended by medical advice" (Guillet, 1954). In the 1840s, the better-funded societies started to build "temperance halls." Small white frame buildings resembling schoolhouses or fundamentalist churches came to dot the rural landscape. These buildings housed the headquarters of the society and were the scene of the speeches, meetings, soirees, and other temperance events.

Early temperance societies allowed the use of wine, beer, and cider and were essentially "anti-liquor" societies. For many years, there was dispute about "the old pledge" (against liquor only) and "the new pledge" (against alcoholic beverages of any sort). For a time there were three types of societies — old, new, and "combined," which had some members who lived under each pledge. The first total abstinence society was formed in 1835 at St. Catharines, and by the early 1840s almost all "old-pledge" societies had disappeared.

Early temperance societies were not populist or democratic organizations. There were numerous racial and other exclusions. For example, blacks and native people were not admitted to societies founded by whites and were expected to establish their own (Barron, 1976). There were a few local black temperance societies. However, native people apparently never founded their own societies. As Lockwood has shown (1993), the temperance movement in Upper Canada was an "ethnic subterfuge." The growth of temperance in the early 1800s coincided with the emigration of Orangemen fleeing Ireland because of rebellions and famines. They often became the heavy drinkers in pioneer communities. Their presence was often disliked by the American emigrants and British upper classes. Many temperance societies were formed by American emigrants to counter the influence of Orange Lodges and to control the drinking of their members, as well as their political and social influence.

For a long time, women were excluded from temperance meetings, as their presence while discussing drink and drunkenness was thought to be unladylike. It was also feared that immorality might result if both sexes went to the same meetings (Guillet, 1954). Eventually, women were admitted as "lady visitors," ineligible to hold office or to be full members. In the long run, of course, women took an active, even dominant, role in the temperance movement, chiefly through their own organizations such as the Women's Christian Temperance Union (WCTU).

During the 1840s a variety of large fraternal temperance organizations were established in Canada (Spence, 1919). The Sons of Temperance, founded by two former drunkards in the United States, began in the Maritimes in 1847. It was very popular because it had the lowest fees and splendid ceremonies, and one could rise through a series of graded ranks with more and more impressive titles and uniforms. Women could join as "daughters of temperance" and children as "cadets of temperance."

The Independent Order of Good Templars, also originating in the United States, came to Canada in 1855. Others, such as the British American Order of Good Templars and the Royal Templars of Temperance, soon followed. They had in common American roots, good organization, devotion to total abstinence, and, at least in the early days, the exclusion of women. They established branches in many towns in the Maritime provinces, linked together through newspapers, exchanges of speakers, annual meetings and the like.

Numerous early temperance organizations were founded by former drunkards, but very few organizations spent much time rehabilitating drunkards or getting them to take the pledge. One of the most interesting of the organizations that did was founded by an American,

David Isaac Kirwin Rine, a former alcoholic and convict. He came to Canada in 1877 and founded the Gospel Temperance Movement after working in a similar organization in Pennsylvania. Unlike most temperance organizers, Rine took a forthright, sometimes crude approach to temperance gatherings and even used "bar-room slang in God's holy edifice." Prayer, speeches, the singing of temperance songs, and pledges were his main methods. To his credit, he focused on derelicts and skid-row alcoholics — those often forgotten by the usual temperance organizations — and charitable institutions. He claimed 190,000 signatures to the abstinence pledge during the five years he worked in Ontario. Of course this number was overstated; it would have constituted about 30 per cent of the adult population of Ontario. Whatever the true figure, we know that Rine gave a very large number of temperance lectures in Ontario and had an enormous impact on the derelict population. However, at the very height of his success, he was accused of indecently assaulting a 15-year-old girl; Rine admitted taking "improper liberties" but denied criminal intent. Nevertheless, his reputation was ruined and his taste for temperance work seemed to disappear. In 1882 he returned to a life of drink and debauchery and died soon afterward, predeceased by the Gospel Temperance Movement.

The early temperance movements must have had a major impact on drinking in Canada. Noel (1994) has reviewed a number of observations from the mid-1800s showing that alcohol consumption among settlers had gone far down. He has also shown that between 1831 and 1861, the number of distilleries fell from 70 to only five in Lower Canada, and from 147 to 53 in Upper Canada. Imports of alcoholic beverages also fell by half, while the cost of whiskey went up by about 50 per cent.

The Women's Christian Temperance Union (WCTU)

The WCTU was a relatively late addition to the temperance movement. It was founded in 1874 at Chatauqua, New York, and in the same year meetings were held in Owen Sound and Picton, Ontario. One of the earliest and best-known WCTU organizers, Letitia Youmans, came from Picton (Cherrington, 1928). The WCTU spread rapidly throughout Ontario and the Maritimes in the 1870s, and by 1885 there were chapters in all provinces. Their motto was "For God, Home and Every Land." Unlike other parts of the temperance movement, the WCTU included blacks, native people and other ethnic groups in its membership. Even men could join, but they did not get voting rights. The best source for material on the WCTU in Canada is the various WCTU yearbooks and R. I. McLean's A *Most Effectual Remedy: Temperance and Prohibition in Alberta* (1969) as well as R. E. Spence's *Prohibition in Canada* (1919).

Although it is best known for its stand on alcohol, the WCTU had a large number of other interests. It placed a great deal of emphasis on women's suffrage and helped to get the vote for women in Canada. Indeed, some of the later WCTU leaders, such as Nellie McClung, were more interested in this aspect of the Union's work than any other. In 1929, four of the five women who took the case declaring women to be "persons" to the Privy Council were WCTU members. Some WCTU members saw women's votes as being necessary to bring Prohibition to Canada, and in that they proved right.

The WCTU also campaigned noisily for observance of the Lord's Day Acts and for feminine purity, but against male infidelity and prostitution, often called the "white slave traffic."* A little known division of the WCTU was even concerned with anti-narcotics activities, the suppression of smoking, and the sale of cigarettes to minors. The "do-everything" approach of the WCTU led some Canadian men to call them the "Women Constantly Tormenting Us" group.

The majority of members in the early WCTU groups were well educated, middle-class, married women with children. Many joined the movement after their families were grown, and others had no families. A large number were teachers, were married to ministers, or were daughters of ministers. Most had a small-town background and were of British ancestry. As Tyrell (1990) has shown, most WCTU members in the 1890s were Scottish and English immigrants who belonged to evangelical churches. There is a striking lack of French-Canadian, other non-British, or Jewish names in the Union's membership lists. Of course, the WCTU had its greatest impact on the "respectable" middle class.

One of the major successes of the WCTU was to have "scientific temperance" established in the schools of English-speaking Canada. The WCTU wanted instruction in temperance to be made compulsory and to have the same status as more traditional subjects, complete with final examinations. As might be expected, given the movement's strength in the region, "scientific temperance" was first introduced

* Numerous books from the temperance movement in general emphasized the connection between sexual excess and drinking. Most were published in the United States. The "Trinity of Evil" referred to in the book of that name by Canon Wilberforce (S. R. Briggs, 1886) included infidelity, impurity, and intemperance.

into Maritime schools, where a course on "physiology and temperance" was taught. However, by 1893, "physiology and temperance" was made compulsory in Ontario schools up to the Grade 8 level. The WCTU financed the preparation of school textbooks on temperance by university professors of physiology. The scientific information in these texts was excellent for the time, and not totally out of date even now. The WCTU's greatest influence in the schools was in the Maritimes, Ontario, and Alberta. In general, the teaching "scientific temperance" tended to follow local option votes; it was much more likely to be accepted in dry areas than wet.

The vast number of temperance organizations and societies soon required some umbrella organization. In 1876 the Dominion Alliance for the Total Suppression of the Liquor Traffic was formed to press the case of temperance with governments and to present a common front. It was organized by Senator A. Vidal, John Cameron, Robert McLean, and Thomas Gales. Ontarians became prominent in the Alliance, and F. S. Spence, of Toronto, acted as its spokesman for many years. The Alliance included representatives from most temperance and Prohibition societies, and constituted, along with the WCTU, the major temperance lobby in Canada for some fifty years.

Social Benefits of the Temperance Movement

There is a tendency to see the temperance movement as a collection of anti-drinking societies, single-minded in their narrow purpose. Although many members also undoubtedly saw them in this way, early societies had a variety of functions and a broad impact on social life.

Early temperance societies were social organizations. They took over many of the foremost functions of taverns, often doing them better and more cheaply. The larger societies put on parties, Saturday night soirées and meetings where politics and local affairs were discussed. There were numerous talks, speeches from visiting temperance leaders, and endless discussions. Societies almost always sponsored bands, sports teams, and concerts of sacred music. In Ontario, nature walks and other outings were also held, and in the mid-1800s, temperance societies frequently sponsored trips to Niagara Falls.

In many centres, temperance societies provided variety and excitement for people who did not wish to socialize exclusively in taverns. In some areas, these societies also provided the only libraries and reading rooms with books, newspapers, and, of course, many temperance tracts.

Temperance societies provided uniforms and regalia for officers and members, flags to carry in parades, and bands to march with. The

Sons of Temperance operated like a secret society with passwords (changed quarterly), insignia for officers, initiation ceremonies, and special recitations and prayers, all meant to emphasize the special status of members (Sons of Temperance, 1876). Many temperance groups also provided fraternal benefits. For example, the Sons of Temperance and the Independent Order of Rechabites used some of their funds to provide benefits during sickness or at death. Temperance workers also had the responsibility of visiting fellow members who were sick. Members had cards and passwords, which gave them introductions to groups in other cities and sometimes access to free accommodation.

A major beneficial effect of the temperance movement was a reduction in social events involving heavy drinking. In areas with temperance societies, alcoholic beverages disappeared from bees, and temperance men refused to go to those that included drink. Also, a variety of dances and other social events became dry. For example, no alcohol was served at the opening of the Welland Canal, where previously much drunkenness would have been expected, because one of the officiating dignitaries was a temperance man, W. H. Merritt (Garland and Talman, 1931).

Temperance hotels also made a brief appearance in Canada, in direct competition with the licensed hotels. Many of these were built in the 1840s and 1850s to give travellers a rest from the noisy, drunk-ridden places so common at the time. They were patronized by temperance-minded people, especially women. However, they were few in number, and some were too stuffy to attract a large clientele. For example, Bible reading and hymn singing often replaced the after-dinner brandy in the parlor. Moreover, the loss of the bar receipts greatly reduced income. Most of them had failed by the end of the 1860s.*

Temperance in Quebec

Since the origins of the temperance movement were predominantly in Protestantism, there can be no surprise that it was mostly unsuccessful in Quebec. Further complicating matters was the fact that the founder of the Quebec temperance movement, Father Chiniquy, ran into many problems with the Catholic Church, eventually proving to be an embarrassment to the movement (Cherrington, 1928). The temperance movement always had difficulties becoming firmly established in

* The reason for the demise of Canadian temperance hotels is worthy of more study. Many such hotels still operate in Scandinavian countries, and some are very good.

Quebec. The official Roman Catholic hierarchy did support moderation on the part of its members. In this century and the last, many Catholic bishops spoke from the pulpit of the need for sensible drinking. However, very few would officially sanction a Methodist-dominated temperance movement. Nonetheless, several useful pamphlets and books about alcohol problems and the need for moderation and even abstinence were published in Quebec in the late 1800s. Those temperance organizations, such as the WCTU and Dominion Alliance, that did organize in Quebec did so later than in the other provinces, and they were successful mainly in the areas where Protestants lived. Even in 1919, the leaders of the temperance movement in Quebec included only people with British (not French) names, according to Spence's *Prohibition in Canada*. Quebec became a headquarters for anti-temperance feelings and the source of much antiprohibitionist sentiment. It was one of only two provinces (with British Columbia) to vote against Prohibition after the First World War.

Father Chiniquy began his temperance work in Beauport, Lower Canada, in 1838 and continued it almost to his death in 1899. Chiniquy linked temperance to French Canadian nationalism and claimed that if the Quebecois did not remain sober, the British and Americans would get all the best land and the jobs. He was a great actor and performer who heightened his talks with stage effects. There were rumors that he arrived "by miracle" rather than by train. Chiniquy's message was simple. It stated that "the national survival of French Canadians depends upon temperance" (Noel, 1993). Most of the heavy drinking in Quebec seemed to be among French speakers and Chiniquy organized campaigns in hundreds of French communities, persuading over 200,000 people to take the pledge against alcohol (Noel, 1993).

Not unexpectedly, Chiniquy's temperance organizing brought him into conflict with the Church hierarchy, and after many disputes he left the Church. He had always had an eye for the ladies and soon married and had children. In 1859, he became an Orangeman and a Presbyterian. He published a number of exposés of life in the Catholic Church. After a period of travel and missionary work, he settled again in Quebec and became one of the most hated men in the province. Nonetheless, Chiniquy remained the unofficial spokesman for the Quebec's temperance movement. Eventually he was officially excommunicated by Rome. People stoned the few Protestant churches in which he was allowed to speak, and it was said that no one would rent him a hotel room in Quebec. Certainly, his activities damaged the temperance movement greatly in that province, although it is doubtful whether the movement could ever have made much permanent progress there, given its close ties to fundamentalist Protestantism and the Ontario middle class.

The long-term effects of Chiniquy are difficult to assess but by 1851 there were abstainers in Quebec, 48 municipalities were dry and Molson's brewery reported losses. It is Noel's contention that Chiniquy "helped to forge the new and lasting image of the church as guardian of the national destiny".

Early Opposition to the Temperance Movement

Naturally, opposition to the temperance movement came from a variety of sources. Drinkers and drink sellers opposed it, as did certain segments of government, which benefited from the alcohol taxes. A number of church groups, especially Anglicans, opposed temperance and favored moderation instead. Many clerics rejected temperance on the basis that it had no biblical support, and numerous ecclesiastical debates were fought on the issue. As the origins of the temperance movement were mainly middle class, it was unacceptable to many in the upper classes. Several temperance leaders became associated with liberal causes in Ontario and were thought to be revolutionaries. However, much of this opposition occurred in the mid-1800s and had largely disappeared by 1900, as the movement grew in strength.

The major anti-temperance forces were drinkers, who wished to continue drinking, and distillers, tavern keepers, and others who wished to continue selling alcoholic beverages. Early temperance meetings were frequently disrupted by hecklers and hooligans (Barron, 1976). Temperance members sometimes had cold water dumped on them. There were also anti-temperance societies of various sorts, especially in Ontario. Bacchanalian Societies were established to continue the tradition of heavy drinking and eating. Hoteliers and distillers helped to create "intemperance" societies, which argued for personal freedom, or "anti-abstinence" societies, which refused to patronize businesses owned by teetotallers.

Naturally, the makers and sellers of alcoholic beverages had support in government and the general population. Many farmers depended upon distillers to buy their grain and feared that runaway temperance would limit such sales. Others benefiting from the drink trade were distillery and brewery workers, the employees of taverns, inns and dram shops, newspaper publishers who took alcohol advertisements, and even the magistrates who gave licences. Both provincial and municipal governments depended greatly on taxes on alcohol produced locally, duties on imported varieties, and licences to operate taverns and stores. Since there were few tax sources in the 1850s, many people in and out of government feared temperance because its success would lead to new taxes.

Alcoholic beverages were also part of the political patronage system. Alcohol was often given away by political parties and candidates at election time, sometimes right at the polls held in hotels. Magistrate positions were frequently given to well-connected distillers and vendors. According to Barron "there was an Upper Canadian magistrate who, as a tavern keeper, sold liquor to his patrons until they got drunk, then issued a warrant for their arrest, tried them on the spot, and finally, in addition to fining them, made them treat one another to settle their differences." Opposition to temperance was prominent among the upper classes in all provinces. In part their stance was related to religion, as most were Anglicans, and in part to the character of the temperance movement. It was very early seen as a lower-middle class movement associated with liberalism. Bishop Strachan, the first Anglican bishop of Toronto, condemned excessive drunkenness but refused all support for temperance societies. His chief concern seemed to be that they were "the mere tool of intriguing politicians, and religious anarchists" (Barron, 1976). In fact, early temperance societies in Ontario did include almost all members of the reform movement who campaigned against the Family Compact. Several of them, such as Marshall Bidwell, Egerton Ryerson, and Jesse Ketchum, were active in the Rebellion of 1837 against the Compact. The charge of "religious anarchy" was also roughly true in that temperance had its origins in evangelical or fundamentalist sects at a time when Anglicanism was essentially the state religion in Upper Canada.

Debate over biblical support for the temperance movement began early and continued throughout its history. A major problem for the movement, especially among religious fundamentalists, was that Christ gave wine to others to drink and drank wine himself. At the feast of Cana, his first miracle was to convert water into wine. Also, there are numerous references in the Bible to the values of alcoholic beverages, although drunkenness is reviled.

The Bible seemed to promote drinking and offer no support to the temperance movement. The Canadian who championed this viewpoint was the Reverend Robert Murray, a Scottish Presbyterian from Oakville, Ontario. He published a series of seven lectures on the topic of alcohol and the Bible and went on extensive speaking tours during the 1830s and 1840s (Murray, 1839). He claimed that teetotalism was against Christ's teachings, and that the manufacture and distribution of alcohol were moral since Christ had done both. Further, he maintained that serving alcohol was part of Christian hospitality, as Solomon had advised that "a feast is made for laughter, and wine maketh merry." Temperance societies, according to Murray, were simply not in keeping with the word of God.

The arguments of Murray and others of the same persuasion were difficult to counter and cost the temperance movement many supporters. A small sub-industry among temperance workers was started to counter the position of Murray. Public debates and discussions were held. Scores of books and tracts were published that purported to show that the wine at Cana was unfermented, that only the Old and not the New Testament allowed drinking of alcohol, and that there was ample biblical support for temperance pledges. Many religious persons remained unconvinced.

The Official Messages of the Temperance Movement

Of course, the first and main message of the temperance movement was that people should not drink. Because the movement took the lead in defining drinking and what should be done about it, anti-drink rhetoric became more complex over time. Early temperance societies were content to gather pledges against drinking and provide a place for non-drinkers to congregate, have fun, and talk about drink. However, after temperance societies turned to government action and Prohibition (see Chapter 3), a variety of messages about drinking appeared. The WCTU and the Dominion Alliance were in the forefront of defining the nature of drink and the possible solutions.

Although the WCTU began as an organization promoting abstinence, it soon saw drink as only one of many social problems. Its basic message was simple — drunkenness leads to poverty, family breakdown, men's infidelity, and hardship for women and children. Get rid of drunkenness and you eliminate all other evils. The WCTU was especially interested in women's vulnerability to men's drunkenness. Married women with drunken husbands and with children to support had no political or legal remedies. Much WCTU literature (for example, the works of Nellie McClung, 1935) portrays men as aggressive, selfish, uncontrollable, and drunk, with women as their victims. McClung often said that "the greatest victims of alcohol had never tasted a drop" — they were the women and children related to alcoholic men.

There were major differences in approach between the Canadian and American women's movements against drinking. Canadian women did not sing hymns or taunt drunkards outside taverns, nor burst in to break the taverns up with hatchets. There was no Canadian Anti-Saloon League or Carrie Nation. As Letitia Youmans said, "We shrink instinctively from notoriety." The Canadian WCTU went to town councils and magistrates who gave licences and petitioned political leaders who set overall policy. The aim was to reduce the numbers of taverns and eventually bring in Prohibition.

The watchwords of the WCTU in Canada were "agitate, legislate and educate," with more emphasis on education. The WCTU strongly supported Prohibition through abstinence pledges and "scientific temperance" education in schools. It also campaigned against sales of alcohol to minors and the illegal trade in alcohol. Even its work with women in prisons emphasized temperance, religion, and right living. It campaigned among the clergy for the use of grape juice or other fruit juice instead of wine at communion services. It abhorred any use of alcoholic beverages at government functions or by government officials. Members vowed to boycott stores that sold alcoholic beverages and tried to get others to do so.

The official position of the Dominion Alliance included the following principles:

(i) That it is neither right nor politic for the state to afford legal protection and sanction to any traffic or system that tends to increase crime, to waste the national resources, to corrupt the social habits, and to destroy the health and lives of the people;

(ii) That the traffic in intoxicating beverages is hostile to the true interests of individuals, and destructive of the order and welfare of society, and ought therefore to be prohibited;

(iii) That the history and results of the past legislation in regard to the liquor traffic abundantly prove that it is impossible satisfactorily to limit or regulate a system so essentially mischievous in its tendencies;

(iv) That no consideration of private gain or public revenue can justify the upholding of a system so utterly wrong in principle, suicidal in policy, and disastrous in results, as the traffic in intoxicating liquors;

(v) That the total prohibition of the liquor traffic is in perfect harmony with the principles of justice and liberty, is not restrictive of legitimate commerce, and is essential to the integrity and stability of government, and the welfare of the community;

(vi) That, rising above sectarian and party considerations, all citizens should combine to procure an enactment prohibiting the manufacture, importation and sale of intoxicating beverages as affording most efficient aid in removing the appalling evils of intemperance (Spence, 1919).

In brief, the Alliance argued that the state cannot afford to allow the sale of alcoholic beverages, since it is morally wrong as well as unwise, and that total Prohibition must be enacted. Their arguments made no special reference to health issues, nor was there any emphasis on banning bars or saloons as there was in the United States. A total Prohibition, nothing less, was demanded.

Eventually, almost everything that went wrong in society was said by temperance workers to be due to alcohol. For example, F. S. Spence claimed that "the common traffic in intoxicating liquors is universally recognized as the most prolific cause of poverty, disease and crime . . . [and] the most formidable obstacle to the moral, social and material progress of our people" (Spence, 1919). The Reverend William Wilberforce (1886) said that "almost all crime, divorce, poverty, prostitution and family problems are due to strong drink." B. H. Spence (the brother of F. S. Spence), like most temperance supporters, took an approach similar to Social Lamarckism, claiming that it was a "fundamental fact" that "alcohol is a narcotic racial poison which damages the individual and blights the race . . . the more of that poison gets into the blood of the people, the less the virility and the more degeneracy there will be in succeeding generations" (Spence, 1925). Temperance workers even claimed that typhoid was at least indirectly caused by drinking, as fewer temperance people than others contracted the disease. Such extreme positions hindered the acceptance of the temperance movement, especially in the 20th century. When Prohibition failed to cure poverty, crime, disease, and mental illness as promised, the expected anti-temperance reaction was not long in coming.

Numerous temperance statements about drinking had no basis in fact. For example, it was often claimed that one drink could turn a man into a drunkard, and that alcohol led directly to insanity. It was also claimed that pauperism was due mainly to drink, and that all those on relief or in workhouses were there because of drink. This last claim could easily have been checked with a little research, but there was no science of human society in the heyday of the temperance movement. A last message was that any who were against temperance were part of "the liquor interests" and not to be trusted.

Another consistent theme was that alcohol had a bad effect on productivity. It was said to promote shirking and to affect the economic output of the whole country. Indeed, there was more temperance emphasis on productivity than on improving the lives of the drinkers. As Barron (1976) has argued, "temperance fed capitalist desires to have a productive and disciplined work force." Drink robbed both the employer and the government by taking time from work. Both puritanism and Methodism held concepts similar to those of the temperance movement about work. As Tawney observed in *Religion and the Rise of Capitalism* (1929), "The good Christian was not wholly dissimilar from the economic man," — and clearly the good temperance man was similar to both. It is no surprise that the supporters of the temperance movement were businessmen, professionals, fundamentalist clerics, and from the middle rather than the lower classes.

The Implied Message of Temperance as Seen through Songs

As well as an official message to governments, "the liquor interests," and the general public, the temperance movement had various implied messages and meanings. It was the main force creating definitions and public attitudes about drinking for nearly a hundred years. This gave its adherents great scope for expounding on all aspects of drinking and drunkenness. Some of the best sources for these expositions are the various collections of temperance songs. Most temperance gatherings involved singing songs about drinking, with most of them set to popular hymn tunes. The first collections of songs came from the United States, but several collections were published in Canada. The most often used collections were Affleck's *Matchless Gems of Stirring Song for Public and Social Gatherings, Schools, Conventions and Camp Meetings* (1877), Whyte's *Nuggets of Gold for Temperance Campaigns* (1898), and *The Canadian Musical Fountain: For Temperance Meetings, Bands of Hope, Temperance Conventions, Social Gatherings, Home Circle etc.* (1874). Through these songs we see the messages about drinking given to new recruits to temperance and the sort of views probably held by most members.

Alcohol was seen as a very powerful substance. One drink could turn you into a drunkard and you were powerless to stop it. Moderate drinking was impossible, and those attempting it were courting disaster.

Say No, to a glass or so,
There's every harm in a glass you know;
And so we've signed the temperance pledge.
"You Must Learn to Say No," *Matchless Gems*

Warn the moderate to beware,
Lest they fall into the snare,
Bid them from temptation fly,
Taste not lest they die.
"Friends of Temperance Onward Go,"
Canadian Musical Fountain

In these songs, all drinking was sinful. As a consequence, drunkards were expected to have miserable lives and to die from drink. They usually died alone, unmourned, in macabre circumstances. The number of such deaths was astronomical and Canadian drunkards often died as a result of winter.

Oh the drinking, sinful drinking,
Glasses ring and voices cheer,
While to drunkards' graves are sinking
Half a million every year.
"Oh the Drinking!" *Nuggets of Gold*

Who killed this man? Was drink his foe?
We found him dead out in the snow.
"Who Killed This Man?" *Nuggets of Gold*

The dreadful meanness of drunkards and their violence toward their families was often the theme of temperance songs. They often beat their wives and families and sent them out of the house — always in winter and usually at the height of a storm.

A drunkard reached his cheerless home,
The storm without was dark and wild,
He forced his weeping wife to roam,
And colder still the winds did blow . . .

At morn her cruel husband passed,
And saw her on her snowy bed,
Her tearful eyes were closed at last,
Her cheek was pale, her spirit fled . . .
"Renounce the Cup." *Canadian Musical Fountain*

Drunkards were seen as the unwilling victims of alcohol, and to have become drunkards without it really being their fault. However, they were able to gain power over alcohol and take the temperance pledge. When they pledged, they were to be accepted warmly into the temperance movement.

> Will you sign the pledge poor drunkard?
> We wish to set you free from appetite and
> passion and custom's slavery,
> Strong drink has been your ruin,
> We ask you to abstain,
> Come throw down the bottle and never drink again.
> "Will you Sign the Pledge?" *Matchless Gems*

> May every drunkard join our band,
> With feelings pure and right,
> And sing while joining heart and hand,
> Good night, good night, good night.
> "Good Night," *Canadian Musical Fountain*

Many temperance songs extolled the great virtues of water. It was seen to be the only acceptable beverage and is the only one mentioned in temperance songs.

> Here's to water pure and bright,
> Sparkling in the crystal light,
> Flashing in the golden way,
> Leaping on its way.
> "Water Pure and Free," *Matchless Gems*

> Oh yes we love the social glass, but it must be filled with water,
> Wisdom says be temperate to every son and daughter.
> "The Social Glass," *Canadian Musical Fountain*

Temperance songs often had a semi-militaristic tone with many references to the "temperance army" and the holy war that would be fought with "King Alcohol." Many songs show a muscular Christianity, sure that the cause is right and God is on their side.

Strong to meet the foe,
Marching on we go,
While our cause we know, must prevail;
Shield and banner bright,
Gleaming in the light;
Battling for the right,
We can never fail.
"Sound the Battle Cry," *Canadian Musical Fountain*

Now the temperance army's marching,
With the Christian's armour on,
Firm and steady is our tread,
Wives and sisters in the throng;
Let the joyful tidings roll.
"The Temperance Army," *Canadian Musical Fountain*

In the battle against alcohol, the main enemies, in addition to drinkers and drunkards, were "the liquor interests" and government and all who stood in the way of Prohibition. Many songs reminded singers of the evil of dealers in alcohol, their perfidy and total degradation. No recovery was possible for dealers in alcohol and they could not join the temperance movement. Governments who delayed Prohibition were merely seen as stupid.

Rum dealers all, both great and small,
Will witness in dismay,
The joyous time when vice and crime,
Shall all be swept away.
"A Better Time," *Canadian Musical Fountain*

In one legislative mill,
to a noted temperance bill,
Men were deaf as posts and blinder than the bats . . .
"Have You Noticed That?" *Nuggets of Gold*

The Decline of the Temperance Movement in the 20th Century

Few people would think of temperance as a 20th-century movement. Most see it as staunchly Victorian. The zenith of the movement probably occurred in Victoria's day, with the national plebiscite on Prohibition in 1898 (see Chapter 3). In the early part of the 20th century the movement was still very strong. However, decline was obvious after the First World War, and even more so after the disappearance of Prohibition. Active membership in the WCTU went from 11,428 in 1911 to 21,042 in 1929, and then declined to about half that in 1939 and only 8,912 by 1950. The number of member unions went from 549 in 1929 to 314 by 1950. The reasons for the decline of temperance are varied, but they involve both the activities of the movement and social changes over which it had little control.

A major factor in the decline was the failure of Prohibition. The movement achieved Prohibition legislation, its major goal, in all provinces except Quebec, only to have it voted out within a few years in all but two provinces. Rejected as unenforceable and unworkable, it was to be replaced by government monopoly systems. With that change, ownership of the drinking question passed slowly from temperance organizations to governments. Various liquor commissions and monopolies began to redefine who could drink, what they could drink, and under what circumstances. As Prohibition disappeared in most provinces during the 1920s, it became clear that it would probably never return. Politicians lost interest, and no major political party seriously proposed it again. As a result of Prohibition's ignominious collapse, the *raison d'être* of temperance disappeared and membership in its organizations fell away precipitously. Other factors in the movement's decline in the 1930s included the Great Depression, which brought reduced incomes and the feeling that there were more important issues than drinking to worry about, together with a relative decline in both drinking and drinking problems.

In the 1920s and 1930s, many temperance activists also showed signs of diversifying their interests. Women such as Nellie McClung left for the more exciting field of women's suffrage, eventually becoming a member of the Alberta Legislature. The WCTU began in the 1920s to be more interested in film censorship and in women's desires for jobs, membership on governing boards of churches and participation in politics. The Manitoba WCTU even took a serious interest in what it considered excessive drinking of Coca-Cola by young people in the 1920s. When universal suffrage was achieved, a major purpose for the WCTU disappeared and no comparable cause emerged to replace it.

To some extent the temperance movement was the passive victim of demographic changes. If we consider the period from 1871 (the first census with data on religion) to 1931 (by which time Prohibition had ended in all provinces except P.E.I.), the changes are remarkable. The number of Anglicans in Canada increased by a factor of three and Roman Catholics by a factor of almost four. However, the religious groups strongest in their support of temperance grew more slowly. The number of Baptists failed to double in the period 1871 to 1931. The Methodists barely doubled in number but were reorganized into the United Church in 1926, where their temperance message was greatly diluted.

Meanwhile, immigration to Canada between 1871 and 1931 involved large numbers of Eastern Europeans and Asians, to whom both Protestant ideology and temperance were totally foreign. Some temperance societies made efforts to campaign among immigrants in the West, but their materials and speakers were rarely suitable. Finally the growing urbanization of Canadian society also undercut the temperance movement. When it was founded, the population was predominantly rural, and the movement always drew its greatest support from rural areas. By 1931 there had taken place an enormous population shift to the cities. In 1871 the population of Canada was 80 per cent rural; by 1931 that figure had fallen to only 46 per cent. People living in large cities were much less likely to support temperance. In general, temperance societies failed to appeal to either immigrants or to the urban lower classes. Although this failure was unimportant in the mid-1800s, it contributed to the downfall of the temperance movement in the 20th century.

The Temperance Movement in the Modern Era

After Prohibition was repealed and governments took over the sale and distribution of alcoholic beverages, the focus of the temperance movement changed greatly. The new enemies were not solely "the liquor interests" but moderationists, government alcohol monopolies, and politicians. From the period of the 1930s, there was a clear increase in the availability of alcoholic beverages. At the close of Prohibition, several provinces required buyers to order alcohol from special dispensaries, with pickup or delivery one or two days later. Eventually the procedure changed so that by the end of the 1930s government stores existed in all provinces except P.E.I. and buyers had immediate access to their purchases.

Under the *Canada Temperance Act* local option votes were still allowed, providing an on-going focus for local temperance activity. At the demise of Prohibition, large areas of Canada, especially rural areas, were technically still dry. Local option votes were required in order that either off-premise stores or on-premise consumption of alcohol or both could be allowed. Temperance forces in all provinces fought local option laws on the dry side, but by the 1950s almost all areas allowed the sale of alcoholic beverages.*

* Much of the material in this section can be found in the archives of the Canadian Temperance Federation and the Ontario Temperance Federation held in the Archives of the United Church of Canada.

The Canadian Temperance Federation and various provincial agencies also attempted to reduce the speed of liberalization in alcohol availability (see Chapter 4), but their efforts were mostly unsuccessful. In all provinces except Ontario, such agencies received small provincial government grants, chiefly for educational work in schools and with young people. Virtually any proposed change in alcohol legislation, whether about advertising, licensing, or opening hours, generated a flood of letters from temperance advocates decrying them. Numerous briefs were presented to provincial governments throughout the 1940s and 1950s on alcohol control policy, but few seemed to have much impact in delaying liberalization. By the end of the 1950s, temperance was a spent concept.

The 1960s brought the near demise of the temperance movement in Canada. In 1967, the activities of the Canadian Temperance Federation and the Ontario Temperance Federation (the largest provincial association) ceased. A consultant's report found that there was no further role for the temperance movement in Ontario. The United Church, which had always been the strongest supporter of the movement, stopped asking its members for abstinence pledges in the early 1960s. The WCTU also continued to decline in importance. By 1960 there were only 6,179 members in 260 unions, half of which were in Ontario.

Today, there are a few hundred members left in the Sons of Temperance in Nova Scotia and a very few in New Brunswick and P.E.I., but apparently none in other provinces. The Federation on Alcohol and Other Drug Problems in Nova Scotia creates an umbrella organization for several temperance groups in the Maritimes.

Currently, there is no national temperance organization except for the WCTU. By the mid-1990s there were only about 600 active WCTU members, together with some elderly inactive members (WCTU, 1994). Since the average age of members is in the 70s, and the death rate consequently very high, the Union may not last much longer. It still holds biennial meetings; the one in 1995 was attended by about 40 members. Despite its dwindling resources, however, the WCTU still takes an interest in general moral reform. Its 1982 meeting included resolutions on such varied matters as battered wives, pornography, lewd television programs, rights of unborn children, and marijuana. There were also resolutions on drunken pilots, drinking on aircraft (by passengers), the need for a maximum .05 blood alcohol level for driving, and indirect advertising of alcoholic beverages. Literature, films, and other educational efforts are still made for the sake of "scientific temperance," and a number of essay contests are held for students.

A few quasi-temperance organizations still exist in Canada. Alcohol and Drug Concerns, Inc. was started in Ontario shortly after the disappearance of the Ontario Temperance Federation. It supports the concept of abstinence but does not require abstinence of members. It does educational work chiefly with young people, church groups, and teachers, and takes an interest in all types of drug use. It receives funds from the United Church and from some 5,000 individual members. Careful attention is paid to avoiding references to "Prohibition," "abstinence," and "teetotal" in its official documents. A similar organization, also with a temperance background, that exists in British Columbia is the Alcohol-Drug Education Service. It, too, receives funds from individuals and churches for alcohol education work.

An Assessment of the Temperance Movement

No social movement in Canada rivals temperance for the length and strength of its influence. Some aspect of temperance became accepted in all provinces, and the movement at one point had influence over the federal and all provincial governments except Quebec. The decline of the movement was typically Canadian in that regional and sectarian variations could not be overcome.

In its early days, the movement provided interesting social and cultural opportunities in a bare landscape. Fun and good times seemed to characterize the early societies. However, after Prohibition became accepted as the sole answer to the drinking problem, much of the fun disappeared and was replaced by earnest political activities, rallies, marches, and confrontations. Ironically, the movement that was originally termed "revolutionary" ended as part of the political and social establishment, branded as conservative and Victorian.

There were many social benefits of the temperance movement that are difficult to assess. It provided an important training ground for many activists, who later went on to other social movements or politics. It also constituted the first real treatment for alcoholics. Numerous reports of drunkards' being cured by joining the movement can be found, although we are not sure of the total number. Of course, the movement brought Prohibition, the impact of which is assessed in the next chapter.

Summary

The temperance movement in Canada began in the 1820s, and few movements can match it for success or longevity. It brought moderation and temperance to a pioneer Canada where heavy drinking was expected behavior in many situations. Early temperance organizations came mostly from the United States and were closely associated with Methodism and other fundamentalist Protestant religious groups. The Women's Christian Temperance Union promoted not only abstinence but a number of other issues, such as women's suffrage and the suppression of smoking, narcotic use, and male infidelity. Unfortunately, the social and cultural benefits of the early temperance movements are forgotten and we remember only their doctrinaire stand against alcohol.

Opposition to the temperance movement came from many quarters, including distillers, brewers, hotel keepers, and some politicians, but also some religious groups who saw no biblical support for temperance. The temperance movement declined because of the failure of Prohibition, but also because of demographic and social changes that reduced the relative size and influence of rural communities and Protestant religious organizations. The WCTU and the remnants of a few temperance organizations still exist in Canada, the latter largely confined to the Maritimes. A few quasi-temperance organizations also exist but, in comparison with the movement in its hey day, they are small and not very influential.

Chapter 3

Prohibition Movements in Canada

Few social movements are viewed as negatively in Canada as Prohibition. From our vantage point in the 1990s, Prohibition is usually seen as a horrible mistake, a temporary aberration in Canadian history, or as "a great experiment that failed." We are likely to believe that the movement was foisted upon Canadians by a small group of religious fanatics, Sunday-school teachers and prudes intent on stopping not only drinking but all other types of fun, too. There is a "thief in the night" feeling about Prohibition perhaps best illustrated in Stephen Leacock's writings (1919). He attributed it to a few greedy, small-minded, small-town shopkeepers, ministers, and farmers who wanted to shut down the gaiety of the city lights. Others speak of the prohibitionist as a "tensely religious type of patriot . . . completely out of character in a setting which had never been noted for its puritanical principles" (Ormsby, 1958). Somehow the opinion has developed that an extreme minority forced Prohibition on an unsuspecting public. That this picture of Prohibition is so far from the truth is due in part to the relative neglect of the era by historians and students of social movements.

We have no sound modern history of Prohibition in Canada, nor is much attention paid to it in standard historical works. Ruth Spence's *Prohibition in Canada* (1919) was written by a prohibitionist and celebrates the prohibitionists and the temperance movement. It was published in 1919 before any of the major Prohibition battles were lost. A few academic theses deal with the subject, but they are mostly unpublished and difficult to obtain (Hallowell, 1972; McLean, 1969). Harold Tuttle Allen's book on the temperance movement in British Columbia does cover early efforts at Prohibition (Allen, 1981). A few journal articles can be found, and a few popular books, such as James Gray's *Booze* (1972) about Prohibition in Manitoba and C. H. Gervais's *Rumrunners* about Ontario (1980). Peter C. Newman's *Bronfman Dynasty: The Rothschilds of the New World* (1978) also includes a colorful account of the roots of the Bronfman fortune in bootlegging during the 1920s and 30s. More recent books by Warsh (1993), Tyrell (1990), Gray

(1982) and others have examined some aspects of Prohibition, contributing to our understanding of the era.

Certainly the leaders of Prohibition movements were sometimes religious and often fanatical. However, the success of the movement depended upon its enormous popular success. There was never a "thief in the night," and Prohibition in Canada was almost always voted in by the electorate. Most provincial and federal elections in Canada after 1860 involved some aspect of Prohibition or stricter alcohol controls. Until 1921, a majority of votes on federal and provincial Prohibition as well as on local options were won by prohibitionists, except in the province of Quebec. Far from being a fraud perpetrated by religious zealots, outside of Quebec, Prohibition was a large-scale national movement.

The majority of active voters in the early part of the century supported Prohibition of various sorts whether local, provincial, or national. However, Prohibition brought unanticipated problems and all provinces eventually rejected it. It is instructive to trace the history of Prohibition movements and to examine the reasons for both their early success and eventual failure.

Early Neo-Prohibition and Local Option Laws

After about 25 years of local organizing, temperance societies in Canada turned to political action to press for Prohibition. In part they were encouraged by the success of temperance forces in Maine, which in 1846 succeeded in having the first state Prohibition introduced into North America. By 1850 forces on the Canadian side of the border were agitating for a "Maine Law" (Spence, 1919). A similar law was enacted in New Brunswick in 1856, but was voted down in Nova Scotia, which retained local option to approve or ban the sale of alcoholic beverages. A chronology of important events in the Prohibition movement is shown in Table 3.1.

In Ontario and Quebec, the *Dunkin Act*, passed in 1864, allowed counties, cities, and townships authority to vote for local Prohibition, but municipal councils could pass such laws without a vote. Any votes were to be verbal ones where the voter said clearly yea or nay in front of his neighbors and friends. Druggists could dispense alcohol "for medicinal purposes," but not between 9:00 p.m. Saturday and 6:00 a.m. Monday. Enforcement of the act was a matter for local authorities.

Struggles to get local Prohibition established occupied the temperance movement throughout the next decade. By 1878 Prohibition votes had been taken in most Ontario counties and townships; temperance

Table 3.1

A Chronology of Events in Canadian Temperance

1807	First temperance society in North America
1827	First temperance society in Canada, Montreal
1838	Start of temperance work in Quebec by Father Chiniquy
1856	First non-wartime provincial Prohibition in Canada, New Brunswick
1864	*Dunkin Act* passed, allowing for local option votes on Prohibition
1874	First Women's Christian Temperance Union meeting at Chatauqua, New York, later in Owen Sound, Ontario
1876	Founding of Dominion Alliance for the Total Suppression of the Liquor Traffic
1879	*Canada Temperance Act* (*Scott Act*) passed, allowing for easier Prohibition votes
1895	Royal Commission on the Liquor Traffic recommends against national Prohibition
1898	Canada-wide plebiscite on Prohibition won, but Prohibition refused by Laurier
1915-17	Prohibition introduced in all provinces because of wartime
1917	National Prohibition imposed under *War Measures Act*
1918	*War Measures Act* ends: Prohibition ends in Quebec but not in other provinces
1921	Prohibition defeated in British Columbia
1923-29	Prohibition defeated in all provinces but Prince Edward Island
1948	Prohibition ends in Prince Edward Island, the last hold-out
1967	Activities of Canadian Temperance Federation and Ontario Temperance Federation cease

forces were successful everywhere except in the larger cities (Toronto, Ottawa, Kingston), a few towns, and a few counties. Very few votes had been taken in Quebec; three counties voted for Prohibition, but the major urban areas did not. Petitions were made for the Dunkin Act to be applied to Manitoba and British Columbia, but nothing came of them, and drinking controls were weak west of Ontario until the late 1880s.

Soon after Confederation, various temperance organizations began agitating for a national local option law based generally on the *Dunkin Act*. Local option laws had been largely unsuccessful outside Ontario and the Maritimes. Even in Ontario a number of areas had repealed *Dunkin Act* prohibitions because of difficulties of enforcement or the campaigns of drinkers and tavern keepers (Spence, 1919).

However, problems arose because the *British North America Act* — the sole constitutional document of the time — failed to mention whether provinces or the federal government had the right to impose Prohibition. In fact, the *BNA Act* did not mention any aspect of alcohol controls except to state that provinces had control over purely local matters and over the licensing of taverns. Prohibitionists agreed that Prohibition fell into the federal powers as a part of the criminal law, or else under trade, commerce, and customs and excise. In fact, various legal wranglings lasted until 1878, when the Supreme Court decided that laws regulating the role of alcohol were in the federal jurisdiction. However, it did not decide whether the federal government could impose Prohibition.

The Supreme Court decision paved the way for parliamentary approval of the *Canada Temperance Act* (known as the *Scott Act*) in 1879. Briefly, the act allowed, but did not require, Prohibition to be introduced by any city or county. A petition by one-fourth of the electors required a poll on Prohibition by secret ballot. A simple majority of voters brought in prohibition of sale of any alcoholic beverages, except for medicinal, sacramental, or industrial purposes. Locally produced wines were not prohibited, nor were they drunk much at the time of the *Scott Act*. In a much amended form, the *Canada Temperance Act* is still on the law books of Canada, but it is virtually never used.

The success of the *Scott Act* in bringing Prohibition varied from one province to another. It was most successful in the Maritimes, where 42 of 47 votes were won by prohibitionists. In Ontario, only 29 of 69 votes were for Prohibition. In Quebec, relatively few votes were taken and only 8 of 17 were for Prohibition (Spence, 1919). Only two votes were held in Manitoba, both in rural areas, and both went for Prohibition. As with the *Dunkin Act*, the larger cities of Quebec, Ontario, and Manitoba tended not to approve of local Prohibition or even require votes. The Act largely served to dry up the Maritime Provinces and some rural areas of Ontario, while leaving most cities and almost all of Quebec unaffected.

It is difficult to evaluate the effectiveness of the *Scott Act*, but at least some data on drinking are available. Table 3.2 shows per capita alcohol consumption for the years 1871 to 1893 in Canada. The general trend from 1879, when the *Scott Act* came into force, until 1893 is certainly downward (from .552 gallons of absolute alcohol to .479), although the difference is not very great. If individual provinces are examined (see Table 3.3), we can see that there was very little change in Ontario and Quebec, which were only partially under Prohibition. There were large increases in consumption in Manitoba and British Columbia, which were not under Prohibition. However, the three Maritime Provinces, most affected by the Act, generally showed decreased consumption. The difference is greatest for Prince Edward Island, where consumption dropped from .278 gallons per capita in 1880 to only .122 gallons by 1893.

The Push for National Prohibition

Clearly, the *Scott Act* had not fulfilled the expectations of prohibitionists. It was ignored in the West and in Quebec and had indifferent results in Ontario. Often dry areas were adjacent to wet areas, and residents who wanted to drink had no problem in visiting them. Enforcement was a constant problem. Cities remained largely outside the Act and provided a Mecca for thirsty rural people. For example, Halifax was the only place in Nova Scotia to get a legal drink. A new round of government action was obviously necessary. The Dominion Alliance was formed, uniting all major temperance groups in an effort to get better federal legislation on Prohibition.

Numerous problems blocked the way. Prohibition was a sensitive problem for all political parties at the federal level, (Brown and Cook, 1974) and none knew how to deal with it. It was a divisive issue with religious, regional, and racial overtones. The Conservative Party appointed a royal commission, which reported in 1895 and recommended against Prohibition, although a minority opinion by a temperance leader was in favor. The Dominion Alliance succeeded in getting a Prohibition plank in the Liberal Party platform and the promise of a plebiscite on the issue if elected. The Liberals won the federal election of 1896 with temperance support and were then required to allow a national plebiscite on Prohibition. Yet it was clear that the Liberal leaders had little taste for Prohibition or the plebiscite.

Table 3.2

Per Capita Consumption of Alcohol in Canada: Gallons of Absolute Alcohol: 1871-1893

1871	.614	**1879**	.552	**1887**	.484
1872	.678	**1880**	.390	**1888**	.456
1873	.681	**1881**	.498	**1889**	.518
1874	.777	**1882**	.557	**1890**	.575
1875	.558	**1883**	.602	**1891**	.495
1876	.607	**1884**	.570	**1892**	.462
1877	.502	**1885**	.612	**1893**	.479
1878	.486	**1886**	.454		

Derived from data in Royal Commission on the Liquor Traffic, Ottawa, 1896.

Table 3.3

Per Capita Consumption of Alcohol in Various Provinces: Gallons of Absolute Alcohol: 1880-1893

	1880	**1885**	**1890**	**1893**
Ontario	.595	.648	.551	.532
Nova Scotia	.273	.272	.255	.244
New Brunswick	.353	.378	.311	.290
Prince Edward Island	.278	.189	.150	.122
Quebec	.532	.630	.574	.529
Manitoba	.337	.764	.531	.540
British Columbia	.759	.806	.870	.996

Derived from data in Royal Commission on the Liquor Traffic, Ottawa, 1896.

Sir Wilfrid Laurier, the party leader, had voted against the Prohibition plank as early as 1887. He was well aware, as a member from a Quebec riding, that the people of that province disliked Prohibition and simply would not have it. The Laurier papers in the Public Archives make it clear that the plebiscite was delayed as long as possible and that only the continual hounding by temperance forces and the Dominion Alliance ever brought it about.* Representatives of all levels of numerous temperance organizations wrote constantly, reminding Laurier of his party's agreement to a plebiscite and inquiring why it was not held immediately. Most of the correspondence on the issue came from Ontario and the Maritimes. Laurier was not particularly concerned with the Prohibition issue, and was far more interested in French language rights, the question of separate schools in Manitoba, and trade and tariffs.

The temperance forces asked that the plebiscite be held "without any side issue" and before the next general election. This was to prove a mistake. The Ontario Licence Holders Protective Association (termed "the liquor interests" by the Alliance) wanted the Prohibition question voted on at the same time as a question on imposing the first income tax. Doubtless they surmised that voters would quickly mark "no" to both questions. Of course, this was not acceptable to the Alliance or to the government. However, the Prohibition plebiscite would have attracted more voters if it had been held at a general election or in connection with some other social policy matter.

Defeated While Victorious: The Plebiscite of 1898

The long-awaited plebiscite was held on September 29, 1898. It asked the rather complex question "Are you in favor of passing an Act prohibiting the importation, manufacture, or sale of spirits, wine, ale, cider, and all other alcoholic liquors for use as a beverage?" Despite good weather in all regions of the country and the pleading of the Dominion Alliance, only 44 per cent of eligible voters turned out to vote. In all provinces except Quebec, there was a majority of votes for Prohibition, although the majority in British Columbia was only 975 votes. There were landslide victories in Ontario and the Maritimes, the temperance strongholds. In Quebec, however, five times as many voted against Prohibition as for it. Out of 543,029 votes cast across the country, 52.5 per cent were in favor of Prohibition. The difference was only 13,687 votes.

---

* Much of the material in this section is derived from the extensive Laurier papers in the Public Archive in Ottawa.

The results of the plebiscite gave Laurier a number of ways to avoid Prohibition. He focused on the poor turnout of voters; 44 per cent was well below the average for general elections. Laurier wrote to F. S. Spence that "only a trifle over one fifth [of the electorate] affirmed their conviction of the principles of Prohibition" and that "the expression of public opinion recorded at the polls in favor of Prohibition did not justify the introduction by the Government of a prohibitory measure." Undoubtedly, Laurier had taken the right decision. If Prohibition had been introduced with so little support, it never could have been enforced in Quebec or British Columbia without massive expenditure and serious social repercussions.

Prohibitionists, of course, felt cheated. They thought they had won only to have crafty politicians snatch victory away at the last moment. They complained about voting irregularities in Quebec (Spence, 1919), and said that the turnout was very good considering that they had no party organization. This last point was untrue, as in most areas they had infiltrated the Liberal Party. Moreover, the prohibitionists had several national organizations, most prominently the Dominion Alliance and the WCTU. B. H. Spence, in letters to Laurier, threatened reprisals at the polls by prohibitionists and claimed that they had been defeated in Quebec by cabinet ministers speaking against Prohibition. He further suggested that Prohibition be introduced everywhere except in Quebec — in fact, that the plebiscite be considered a large-scale "local option" vote. Laurier, though, would have none of it and national Prohibition was shelved until the First World War.

The prohibitionists lost the fight for several reasons. They did not get out the vote, although conditions were favorable to do so. They insisted that the plebiscite not be associated with any other issue but, as the turn out showed, Prohibition was not of sufficient interest on its own. Their organization in Quebec had always been weak, dependent almost entirely on English-speaking Quebeckers and the few Protestant churches. The Catholic churches and priests spoke frequently about the need for moderate drinking but would have nothing to do with temperance organizations run by Methodists and Baptists from Ontario (Brown and Cook, 1974). Also, prohibitionists had neglected to get from Laurier a clear statement that a simple majority of those voting, rather than of the total electorate, was sufficient to bring in Prohibition. They never made that mistake again in dealing with governments.

Prohibition During the First World War

After the failure of the national plebiscite, the Dominion Alliance and other temperance organizations turned their attention to the provinces. Efforts were made to have Prohibition introduced in all provinces, but only in P.E.I. did they succeed (in 1902). Local option laws were greatly strengthened in a few places (Spence, 1919) but national Prohibition ceased to be a realistic goal for many years after 1898.

The First World War brightened the outlook for prohibitionists. It allowed them to shift from insisting on Prohibition as a cure for violence, poverty, privation and hunger — a claim that few seemed to believe — to insisting that supporting it would be an act of patriotism. Repeatedly, prohibitionist groups claimed that Prohibition would help win the war by improving military and industrial efficiency. Grain and fruits used in beverages could better be used to provide food for the population. The Dominion Alliance resolved in 1916 that "in view of the necessity for conserving our country's national resources and preventing any impairment of the efficiency of our country's manhood . . . the Dominion Government and Parliament be earnestly urged to enact as a war measure, a law prohibiting the manufacture of intoxicating liquor for beverage purposes" (Spence, 1919). In 1916 temperance organizations ran an advertisement that said:

> Are We To Do Our Duty By the Empire Or Are We To Neglect It?
>
> Are we to "Be British" indeed and remove a "greater enemy than the Hun" from our midst? Is the sacrifice made by our soldiers for us on the battle field to be the only sacrifice? The Bar or the War? That is the Question of the Hour (Hiebert, 1969).

The British themselves never introduced Prohibition, nor did any protagonist other than the United States, Canada and Russia. All available evidence suggested that soldiers hated Prohibition and voted against it whenever possible. Nevertheless, all Canadian provinces approved Prohibition in 1915 or 1916, with Quebec following in 1917. At the end of 1917, Prime Minister Robert Borden announced national Prohibition without a plebiscite. Beverages with more than 2.5 per cent alcohol were not to be sold, manufactured, imported, or distributed. Borden's statement alluded to the need for a "vigorous prosecution of the war" but made no reference to the traditional temperance attitudes about alcohol. Without the war, it is very doubtful that national Prohibition would ever have come to Canada.

One further, little-known aspect of Prohibition was the efforts of the WCTU and other temperance groups to impose dry canteens on Canadian soldiers in Europe. The Dominion Alliance had always distrusted foreigners as being both confirmed drinkers and hostile to Prohibition. Several of their resolutions argued for Prohibition on the basis that Canada would soon be corrupted by "the tremendous tide of immigration pouring in amongst us from the countries where the liquor traffic is not so effectually curbed as here" (Spence, 1919). The National Council of Women said that they were willing to send their sons abroad to fight the war but "with acute pain that they have learned that in so doing they have not only imperilled the lives of their loved ones, but in addition have subjected them to influences, the result of which may be even more deadly than shrapnel or shell" (Brown and Cook, 1974). The WCTU collected 66,186 names on a petition to the British War Office not to allow Canadian soldiers into wet canteens. They also protested the distribution of free rum to soldiers and sailors, and even attempted to stop the shipment of cigarettes to the trenches (King, 1942). While the WCTU was not successful in having alcoholic beverages removed from canteens, returned soldiers later disrupted WCTU Tag Days and retained a strong dislike for prohibitionists in any form.

Postwar Prohibition: The Successful End of a Long Fight

At the end of 1919, the federal government allowed the *War Measures Act* to lapse and with it the regulations for Prohibition. Provinces fielded referenda in 1920 or 1921 to elicit public opinion on continuing Prohibition. In Ontario, Nova Scotia, Saskatchewan, Alberta, Manitoba, and New Brunswick, Prohibition continued into the 1920s, and in the case of P.E.I. well beyond, to 1948. However, it was defeated in 1921 in British Columbia. In Quebec, beer, wine, and cider (but not spirits) were allowed in 1919. Prohibition had never been popular in Quebec. In British Columbia, it had minimal support and numerous critics (Hiebert, 1969).

Prohibition lasted varying lengths of time in the Canadian provinces (see Table 3.4) and was introduced after the war for various reasons. In all provinces, the Dominion Alliance and the WCTU fought tough campaigns organized by experienced workers. The electorate were essentially voting for a continuation of a set of controls to which they had become accustomed during the war. Also, the postwar Prohibition plebiscites allowed women to vote — for many, for the first time. Women's organizations, such as the national Voice of

Table 3.4

Dates of Prohibition and its Repeal in Canadian Provinces

Saskatchewan	1916-1925
Alberta	1916-1924
Ontario	1916-1923
British Columbia	1917-1921
Quebec	1918-1919
Manitoba	1916-1924
New Brunswick	1917-1927
Nova Scotia	1916-1929
Prince Edward Island	1907-1948

Women and the WCTU, were greatly in favor of Prohibition, as were most women.

A further factor is that Prohibition in many provinces became part of the "social gospel." Prohibitionists successfully convinced the population of many provinces that society was being harmed by alcohol and that it was the duty of all Christians to support its removal (Hallowell, 1972; Thompson, 1969; Forbes, 1971). By the end of the war, almost all major Christian denominations had taken some official stand on Prohibition. The traditional stand of Methodists, Baptists and Presbyterians was, of course, to favor it. However, in the Maritime Provinces, even the Anglican and Catholic churches officially supported Prohibition in the 1920s. In its widest form, the social gospel deemed that all sorts of earthly evils could be defeated, including drinking, smoking, opium use, gambling, and prize fighting, as well as vice and corruption. There was great hope that legislation could make a better society, which Prohibition would help to usher in. The social gospel message was that legislation could bring "the kingdom of God on earth."

It has also been pointed out that the 1920s was the "era of prohibitions." In Ontario in 1919, for example, one could not buy a newspaper, a loaf of bread or an ice cream cone on Sundays and it was illegal to play baseball, tennis or golf under the *Lord's Day Act*. There were no buses or streetcars running on Sundays. Movies and magazines were heavily censored. It must have seemed natural to many that there were no alcoholic beverages for sale and no taverns.

Problems with Prohibition

Almost as soon as it was introduced in most provinces, numerous problems arose with Prohibition. It seemed to lose support very quickly among both politicians and the general public. In Quebec, alcoholic beverages were legal, and it was noted in several provinces that Quebec had a boom in the tourist trade and that the government there gained millions of dollars from taxes on alcoholic beverages and licences to sell them (Brown and Cook, 1974). Calls for repeal and government control of alcoholic beverages on a "Quebec model" began almost immediately.

Other problems with loopholes and enforcement arose almost as quickly. There were numerous legal ways to get around the Prohibition system. Of course, most bars and taverns were successfully banned, but drinking was not. Breweries remained to make beer "for export" or near beer, called temperance beer. Near beer (2.5 per cent alcohol) was sold extensively in British Columbia, Saskatchewan and Manitoba, as were "native" (locally produced) wines in Ontario. In western provinces, alcoholic beverages could be imported from other provinces for at least the first few years of Prohibition.

Provinces had no right to interfere with interprovincial trade, but eventually federal legislation prohibited interprovincial trade in alcohol. Distillers in Ontario and Quebec established depots in towns near provincial borders (Kenora, Yorkton), and residents of other provinces could legally order liquor from them. In Ontario up to 100 bottles of liquor could be ordered in Quebec and shipped or carried to Ontario. Much liquor made in Ontario was supposedly shipped to Cuba or to Mexico, but actually went to the United States or to bootleggers in Ontario or other provinces (Gervais, 1980). For a long time Canada was a supplier to the thirsty citizens of the United States suffering under Prohibition. Many small shipping companies existed by shipping Canadian-made alcoholic beverages to the United States or other parts of Canada. For example, the Mexico Export Company had five docks in Windsor but nothing larger than a small motor boat. One motor boat filled with liquor is recorded in customs papers as having made four trips from Windsor to Cuba in a single day. Provided they could order ahead and pay in advance, wealthier drinkers had no reason to be short of alcoholic beverages. Many immigrant families made their own liquor or wine. Convictions for illicit stills went up remarkably during Prohibition. Saskatchewan had 60 per cent of the convictions for illicit stills in Canada, and it was estimated that there were 200,000 convictions in the province in 1921.

Another loophole was the prescription of alcoholic beverages by physicians.* This seems to have been legal under all Prohibition laws, and it represented an important boon to both drinkers and enterprising physicians. Alcohol had been frequently prescribed before Prohibition for heart conditions, anxiety, sleeplessness, and a variety of other complaints. It was viewed as legitimate medicine at the time by both doctors and patients. Even veterinarians in Saskatchewan fought for the right to prescribe alcohol, and soon received a monthly quota of two bottles (Pinno, 1971).

---

* The first author's grandfather, a Methodist preacher and a temperance worker, was first prescribed brandy for his heart condition about this time. He took it regularly and never considered it "drinking" or a violation of his temperance vows.

Prohibition resulted in floods of prescriptions for alcohol, and soon governments were selling far more alcohol, through physicians, than any bootlegger. By 1923, the Ontario government had sold nearly $5 million worth of alcohol by prescription. In 1923-24 some 810,000 prescriptions were issued. One doctor in Ontario had given 2,005 in a month and received two or three dollars for each; another gave 487 in a single day (Hallowell, 1972). In Saskatchewan, pressure on the government was so great that for a short time pharmacists were allowed to dispense alcohol without a prescription — 8 oz. to a person living within five miles, 16 oz. to those living further away (Pinno, 1971). In British Columbia, one physician ordered more than 4,000 bottles of alcohol in a month, and several had more than 2,000. In Manitoba, a doctor and pharmacist combined forces to issue 4,800 alcohol prescriptions in a single month (Thompson, 1969).

As Stephen Leacock wrote in 1919: "It is necessary to go to a drug store and lean up against the counter and make a gurgling sigh like apoplexy. One often sees these apoplexy cases lined up four deep." Long line-ups outside drug stores were frequently found, especially at Christmas and other public holidays. In Prince Albert, police claimed that most cases of public drunkenness and brawls involved prescription alcohol. Efforts were made to limit the number of prescriptions that doctors could issue, but they were never very successful. However, some 311 physicians in Ontario had their privileges suspended, as did smaller numbers in other provinces.

Numerous problems with enforcing Prohibition occurred early and worsened as time passed. Much of the enforcement depended upon local police and customs officers, who had many other tasks; they were also subject to bribery by rumrunners and owners of illicit taverns. Special agents were hired who carried guns and used open search warrants. In Ontario and Saskatchewan, authorities resorted to the use of "spotters" or "stool-pigeons" who spied on the general public, likely bootleggers, and speakeasy operators. Because so many of the spotters were thugs and petty criminals who were later corrupted, the practice brought enforcement of Prohibition into disrepute. Also, beer and liquor were legally made in Ontario (six distilleries and 29 breweries) for shipment to the United States, which was also under Prohibition. Such shipments through Detroit and Buffalo were legal under Canadian law but, of course, not under United States law. Public opinion found an inconsistency in this arrangement.

Bootlegging also continued in all provinces under Prohibition. Whole trainloads of bootleg liquor were seized in several provinces.

Small amounts were seized daily. Convictions under the liquor control acts went up greatly everywhere (see Table 3.5). It was soon obvious to everyone that neither bootleggers nor illicit taverns had been closed down. For example, Gervais could describe several dozen in the Windsor area alone (Gervais, 1980). Many large hotels were convicted over and over for selling illegal spirits and beer.

In several provinces, the officials and special agents enforcing Prohibition ran into trouble. In Ontario, J.O.L. Spracklin, a former Methodist minister, was appointed special agent in the Windsor area. His men carried guns, clubs, and open warrants (writs of assistance). In the course of an investigation, he shot and killed a well-known bootlegger — supposedly in self-defence. However, public opinion tended to see this killing as over-enforcement, and there was a negative view of special agents carrying guns. Similar problems with corruption occurred in the West. In Saskatchewan, R.E.A. Leech, the Chairman of the Liquor Commission and a former temperance leader, was often accused of selling seized liquor to bootleggers. He lost a libel suit against the *Regina Leader* newspaper on the issue. In British Columbia, the first Prohibition commissioner, W.C. Findlay, a former temperance worker, became a prominent bootlegger while in office. He spent several years in the penitentiary, but several of his associates escaped to the United States. Corruption and dishonesty among the enforcers of Prohibition was a constant problem. In addition, the Wild West flavor of Prohibition enforcement bothered many Canadians.

The Reaction against Prohibition

After most provinces had experienced Prohibition for a few years, public support for it waned. Prohibitionists had always failed to appeal to city dwellers, immigrants, and the working classes. These segments of society were greatly increasing, whereas the numbers of Methodists, Baptists, and confirmed temperance followers were not. Loopholes, failures of enforcement, and the draconian methods used by police and special agents seemed to turn public opinion. It was also clear that Prohibition had not brought the great society promised. Poverty, crime, disease, and economic misery still remained. The post-war years were a time in which there was disillusionment about the chances for moral progress in Canadian society through social legislation (Hiebert, 1969). Belief that the social gospel could bring the kingdom of heaven on earth evaporated.

Support for Prohibition waned even among legislators. For example, the "last-night party," or end-of-session party, held by the

Table 3.5

Mean Rates Per 100,000 Aged 15 Years and Older for Drunkenness and *Liquor Control Act* Violations

	DRUNKENNESS	L.C.A. VIOLATIONS
Saskatchewan		
Pre 5 years 1911-1915	628	105
Prohibition 1916-1925	167	164
Post 5 years 1926-1930	132	242
Alberta		
Pre 5 years 1911-1915	1873	190
Prohibition 1916-1924	364	219
Post 5 years 1925-1929	352	205
Ontario		
Pre 5 years 1911-1915	759	112
Prohibition 1916-1923	564	176
Post 5 years 1924-1928	611	261
British Columbia		
Pre 5 years 1912-1916	2112	143
Prohibition 1917-1921	529	267
Post 5 years 1922-1926	381	337
Quebec		
Pre 5 years 1913-1917	744	72
Prohibition 1918-1919	504	96
Post 5 years 1920-1924	559	102
Manitoba		
Pre 5 years 1911-1915	1875	35
Prohibition 1916-1924	470	89
Post 5 years 1925-1929	439	181

United Farmers Organization in Ontario in 1921 had both women and liquor (Hallowell, 1972). It was held in the legislative buildings by a political party elected on a temperance platform. (Even worse, some of the women present were said to be from France!) Numerous politicians declared Prohibition to be inherently unenforceable years before it was repealed.

It was obvious by 1922 that both British Columbia and Quebec had systems of government controls in operation that avoided both the excesses of Prohibition and the unchecked sale of alcoholic beverages seen in the 1880s. These systems helped governments that were short of money to find a new tax base, and were welcomed for that reason. In Nova Scotia, it was clear that repeal of Prohibition allowed the government the only way to finance participation in a new pension scheme. Nova Scotia politicians noted that government profits in Quebec from selling alcohol were close to their own total provincial budget.

Moderation leagues developed in the western provinces, especially in Manitoba, British Columbia, and Saskatchewan. These were well-organized societies of moderates rather than "the liquor interests." Their arguments for government control of alcohol eventually gained great public support. Moderation leagues carefully cultivated an image of respectability; many of the leaders were former army officers and prominent businessmen. Although branded by temperance workers as "the liquor interests in sheep's clothing," connections with such interests could not often be proved. League members pointed out, quite rightly, that they, and not prohibitionists, were the champions of true temperance. They also argued that Prohibition was un-British and "Yankee," and that having it would discourage "the right kind of immigrant" in Canada. The leagues suggested that governments should take over the sale of alcoholic beverages and that the resulting tax revenue should be shared with municipalities and used to offset provincial debts. They were not in favor of the return of saloons. Moderationists pointed out that Prohibition discriminated against poor people who were unable to order beverages in bulk from other provinces. Of course, they repeated what everyone knew — that Prohibition laws were not enforceable. In general, moderationists argued persuasively for an end to Prohibition, and much of what they said was both accurate and timely. They succeeded in British Columbia, Saskatchewan, and Manitoba in getting referenda on Prohibition, which they won. Surprisingly, temperance forces largely ignored the moderationists and greatly underestimated their strength.

By 1929, all provinces except P.E.I. had voted for repeal of Prohibition, but none returned to the wide-open days of the 1880s. Instead, they developed government control systems in which the government was the sole supplier of alcoholic beverages, whether to individuals or to bars and taverns. In some provinces, government stores were established, while in others alcohol was ordered from dispensaries for later delivery. There was no return to the heyday of the bar and saloon or to wide-open drinking.

The Impact of Prohibition on Drinking and Drinking Problems

So far we have focused on the social and political impact of Prohibition. In general, the movement was a social and political failure in Canada, as in all cases (except P.E.I.) Prohibition was repealed within a few years. However, the question remains as to its impact on drinking and related problems. Of course, Prohibition was far from perfect. It limited access to alcoholic beverages, but certainly did not do away with them, as we have seen. What the effects of total Prohibition, without doctors' prescriptions, imported alcohol, or bootleggers, might have been is a matter for speculation.

Unfortunately, we do not have good data on just how much alcohol was drunk during Prohibition, since so much was illegally or secretly produced. However, a variety of other data do exist. We have statistics for several provinces on convictions for drunkenness and convictions under the provincial *Liquor Control Acts*. These data are shown in Table 3.5 for the five years prior to Prohibition, the years of Prohibition, and the five years afterward. The provinces shown are all those other than the Maritimes. The latter are left out because of their small populations, and because in two (Nova Scotia and P.E.I.) Prohibition lasted so long. The data show that in all provinces, convictions under the *Liquor Control Ac*t went up during Prohibition, sometimes by a factor of two or more. In all provinces other than Alberta, such convictions continued to increase after Prohibition. (Note that offences under these acts would, of course, be differently defined in each period.)

Data for public drunkenness show a different pattern. Especially in western provinces, there were great reductions in drunkenness convictions during Prohibition years. The change was remarkable at the time and obvious to many observers, who could see drunkenness suddenly disappearing from the streets. After Prohibition, drunkenness figures stayed down and in most provinces actually decreased rather than returning to pre-Prohibition levels. In general, then, Prohibition seems to have had its greatest effects on public drunkenness in western provinces. Since government sale of alcoholic beverages succeeded Prohibition, a return to pre-Prohibition levels would not have been expected. Also, because of the economic depression in the late 1920s and early 1930s, the buying power of the public had greatly declined.

Rates of deaths from alcoholism and liver cirrhosis are available prior to 1921 for Ontario, British Columbia, and Quebec (see Table 3.6). In Ontario, the number of cirrhosis deaths during Prohibition fell to only 40 per cent of the figure for the previous five years. In British Columbia, cirrhosis deaths fell to 35 per cent of the pre-Prohibition level. In Quebec the reduction was smaller. In both Ontario and British Columbia deaths from alcoholism increased after Prohibition. In general, Prohibition was associated with large reductions in deaths from alcoholism. Rates of deaths from liver cirrhosis — a good indicator of alcohol problems — show a roughly similar pattern to that for deaths from alcoholism.

Because Prohibition was not total and the post-Prohibition era involved government control of access to alcohol, the two periods may not be as completely different as we would at first imagine. While rates of drunkenness and alcohol-related death decreased during Prohibition, in most cases the rates stayed down afterward. Prohibition's greatest impact appears to have been on public drunkenness. Thus, the generalization that Prohibition "failed," needs to be qualified. It is important to specify in what ways it failed, and to indicate the ways in which it was successful.

Does Prohibition Have a Future in Canada?

Our tendency now is to see Prohibition as an outmoded social policy, relegated to the pages of history. There can be little doubt that Prohibition could not return as a general policy under normal circumstances. Support for temperance movements has disappeared, and public attitudes toward alcohol have become extremely permissive in recent decades. Under unusual circumstances, a prohibition of a limited sort may be reintroduced. Highly restrictive controls on drinking were introduced by the federal government during the Second World War (King, 1942). Distilling was eventually forbidden, and there were large reductions in amounts of alcohol sold and in hours of sale. Had the war continued much longer, total Prohibition might well have been reintroduced.

Another setting for Prohibition measures has been certain Indian reserves, particularly in the North. Several reserves introduced Prohibition in the late 1970s and found, as with the experience 50 years earlier, a reduction in drunkenness but problems with enforcement (Smart,

Table 3.6

Mean Rates of Alcohol-related Deaths Per 100,000 Aged 20 Years and Older

	Alcoholism	Liver Cirrhosis
Ontario		
Pre 5 years 1911-1915	6.5	7.0
Prohibition 1916-1923	2.8	5.2
Post 5 years 1924-1928	3.9	5.3
British Columbia		
Pre 5 years 1912-1916	11.8	5.7
Prohibition 1917-1921	3.9	4.4
Post 5 years 1922-1926	6.1	6.1
Quebec		
Pre 5 years 1913-1917	4.6	14.2
Prohibition 1918-1919	4.1	13.2
Post 5 years 1920-1924	1.8	10.8

1979a). Limited Prohibition to solve particular local problems of heavy alcohol use is an experiment that may be worth repeating in many parts of Canada.

An Overall Evaluation of Prohibition as a Social Measure

Prohibition in Canada represented the culmination of nearly a hundred years of effort by innumerable temperance groups and countless Canadians. Its failure was a blow from which the temperance movement never fully recovered. It was a social policy failure that led to corruption and dishonesty among police and enforcement agencies on a scale never seen before. Disrespect for the law among bootleggers, rumrunners, distillers, and brewers was obvious. However, there was also the disrespect shown by the thousands of citizens who bought alcohol on prescription or frequented the numerous "blind pigs," speakeasies, or underground taverns. There were also the many thousands who operated illicit stills or bought alcohol illegally. Disappointment with Prohibition began almost as soon as it came into force, and many ways were found it get around it.

Certainly the reality of Prohibition was a disappointment to temperance leaders, who constantly campaigned for stricter laws and better enforcement. However, if laws had been stricter and enforcement better, the result probably would have been more corruption, violence, and lawlessness. It is very doubtful that the people of Canada would have tolerated more stringent controls. The loopholes allowed some alcoholic beverages to be provided to those who wanted them.

Ironically, while temperance leaders often pointed to the reduction in drunkenness during Prohibition, they made little use of the data showing it to be a successful public health measure. That Prohibition had important effects on health was of no interest in the 1920s. Now that its public health benefits have been established, Prohibition is no longer of interest.

Prohibition by popular vote came too late to Canada. The heyday of prohibitionist feeling occurred before or during the First World War. Two provinces, British Columbia and Quebec, did not continue postwar Prohibition for long, and their control systems provided a competing ideology for Prohibition, as well as taxes in abundance. Moreover, by the time Prohibition referenda were introduced after the war, some temperance leaders and groups had moved on to other things, such as suffrage and women's rights. Also, the demography had changed so that Canada was no longer a predominantly rural country with a strong Protestant influence. The new immigrations of the 1920s were not primarily from countries with temperance traditions, and few immigrants were Methodists. That they had little use for Prohibition was demonstrated at the polls.

Summary

Prohibition is often viewed negatively in Canada. Frequently, it is seen as a social mistake, a failed experiment, or the result of pressure tactics by a few religious fanatics. In fact, it was a very popular movement in Canada up until the 1920s. Temperance movements began agitating for Prohibition in the 1850s after it was achieved in Maine. The first vote on Prohibition was won in New Brunswick in 1856. There followed a succession of laws allowing local option votes. Most of these, outside of Quebec and a few large cities, were won by the prohibitionists, and a large part of the country was legally dry by the mid-1890s. A national vote on Prohibition was won by a small margin, but Prime Minister Laurier declined to bring in Prohibition because of the small size of the majority and the low voter turnout.

Prohibition came to all provinces during the First World War, but was defeated in various plebiscites in the 1920s. Surprisingly, it lost support with both politicians and the general public soon after the war. Enormous problems of legal enforcement arose, and it was obvious that many people still wished to drink. Alcohol was available on prescription to almost anyone. It could also be imported from other provinces in the early days of Prohibition. Bootleggers, speakeasies, and other illegal sources supplied a bustling underground market. Although Prohibition must be seen as a social and legal failure, it did reduce rates of drunkenness and deaths from alcoholism and liver cirrhosis by large margins. Prohibition's failure was a blow from which the temperance movement never recovered. Ownership of "the alcohol problem" passed from temperance movements to governments, who brought in limited access to alcoholic beverages through provincially controlled commissions.

Chapter 4

Alcohol Consumption in Canada Today: Who Drinks, How Much, and Where?

At present, most people in Canada drink alcoholic beverages at least occasionally. Most drink infrequently and prefer small amounts, but many do otherwise. Although never one of the world leaders in alcohol consumption, Canada has retained a respectable position in the top quarter of consuming nations. Today, we have a large amount of information on how much alcohol is drunk, how it is drunk, and in what form. Most of this information is relatively recent, with the first surveys appearing in the 1940s. However, some estimates of overall per capita consumption can be found as far back as 1871 (see Chapter 3). We are concerned in this chapter with drinking in the general population; specific groups at especially high risk of heavy drinking will be reviewed in Chapters 5 and 6.

Methods of Measuring Drinking

Among alcohol researchers, much rancorous debate occurs about the best methods of assessing drinking (Popham and Schmidt, 1981; Lemmens, 1994). Many people prefer surveys of general or special populations. Others criticize surveys as unreliable and accept only per capita consumption figures from government sources. Still others prefer observational studies in which drinkers are actually observed doing their drinking in bars, taverns, or party situations. All methods have something to offer, but we should also look at the limitations of the various methods.

One of the most frequently used methods for studying drinking involves surveys of the general population. Such surveys usually consist of household interviews in which one or more persons is asked about his or her drinking. Detailed questionnaires have been developed to elicit information about all aspects of drinking — how much is consumed, how fast, where it occurs, with whom, under what emotional conditions and so on. Several problems with drinking surveys have become clear. In Canada, surveys cannot account for more than about 40 to 50 per cent of the alcoholic beverages sold (Pernanen, 1974). A good example of this is the *National Alcohol and Other Drugs Survey* (NADS) done in 1989 (Eliany, Giesbrecht et al., 1992). When we look at the estimated number of drinks per adult per week from sales data, it varies from 11.5 in British Columbia to 7.5 in New Brunswick. According to the NADS data, however, the reported number of drinks per week was between 2.8 and 3.9. In all provinces the national survey underestimated the amount of alcohol sold by 50 per cent or more (see Table 4.1). Also, various methods of measuring drinking produce somewhat different results. For example, questions anchoring alcohol use in specific social situations produced higher consumption figures than the more commonly used quantity-frequency questions (Single and Wortley, 1994).

Considerable work has recently been done to improve alcohol-use surveys. For example, Lemmens (1994) has shown that surveys underestimate drinking because in social situations people serve drinks larger than the standard ones and he estimated that surveys under-report real drinking by about seven per cent. Other research (Room, 1990) has indicated that asking questions about recent drinking, rather than lifetime drinking or drinking in the past year, gives the most valid results.

Household surveys, of course, miss many people who might be heavy drinkers (e.g., those in the army, in prisons or hospitals, or in treatment centres for alcoholism, as well as university students in residence, transients, and people living on skid row). Some studies have shown that drinking surveys are distorted by under-reporting by heavy drinkers (Popham and Schmidt, 1981). While other studies have not identified this as a problem (Boland and Roizen, 1973), in general, it appears likely that there is some under-reporting by heavy drinkers in alcohol surveys. Of course, surveys do identify many heavy drinkers and many people with alcohol problems. For example, a recent survey in Ontario (Adlaf, Ivis and Smart, 1994) found that 15.6 per cent of drinkers reported a significant level of alcohol problems — about the number expected according to some studies of sales data. Most drinkers in alcohol surveys probably report their consumption reasonably well — certainly well enough to classify them as heavy, medium, or light drinkers and to study their characteristics.

Alcohol surveys have limitations for studying the total amount of alcohol consumed, but they are one of the few ways to study the characteristics of users. Certainly they can be improved. Efforts have

Table 4.1

Estimated Number of Drinks Per Week in Canada, Ages 15 and Over from Sales Figures and the 1989 *National Alcohol and Other Drugs Survey*

Province	1980-81	1985-86	1990-91	1989 Survey
Nfld.	12.2	10.1	10.4	3.7
P.E.I.	11.7	9.7	8.5	3.1
N.S.	11.5	10.3	9.2	2.8
N.B.	10.4	8.8	7.5	2.8
Que.	11.3	10.0	8.8	3.9
Ont.	12.9	11.8	10.5	3.9
Man.	12.8	11.2	9.1	3.0
Sask.	11.5	10.2	8.4	3.4
Alta.	11.7	13.0	10.7	3.7
B.C.	14.8	12.9	11.5	3.8
Canada	12.4	11.3	10.0	3.7

been made in Canada to use an "informant" method for alcohol surveys in which people are asked not about their own drinking but about drinking in their occupational group (Smart and Liban, 1982). This method is more successful than direct surveys, in that it accounts for about 80 per cent of the alcohol sold.

Many alcohol researchers rely on government sales data as valid measures of alcohol consumption. Such data are the basis of all "per capita consumption" figures. However, per capita consumption figures have only a tenuous relationship to the actual *pattern* of consumption. The figures are based on stocks of beverages moved from warehouses to retail stores. There are seasonal variations in these movements that do not reflect actual drinking. For example, it is well before Christmas that large stocks are moved in anticipation of heavy buying and drinking at Christmas (Lemmens, 1994). Per capita consumption figures also fail to take account of tourism. For provinces such as Ontario, it has been estimated that tourism has little impact on total consumption, as visits in and out of the province are about equal. However, tourists' visits to Prince Edward Island, for example, would far exceed residents' visits elsewhere. Also, Canadians are among the most frequent travellers in the world, but per capita figures do not account for alcohol bought in duty free shops or brought into the country legally and illegally by Canadians travelling abroad.

Recently concern has been expressed in several provinces about an increase in alcohol smuggling, mostly across the border with the United States. In New York, for example, distilled spirits are about half the cost in Ontario; beer and wine are also cheaper but not to the same extent. A number of smugglers have been convicted for bringing hundreds of cases of spirits across Lake Ontario in small boats. There is speculation that much of that alcohol would be sold to bars where it would be put into mixed drinks without the customer knowing where it originated. A number of charges have been laid against bar owners and the Liquor Control Board of Ontario (LCBO) conducts special investigations for contraband alcohol. In 1994, the LCBO estimated that 15 per cent of the Ontario beverage market was made up by smuggled alcohol. If that is correct then official sales figures seriously under-represent how much people are drinking.

Per capita consumption figures from sales cannot take account of home production of beer and wine. It has been estimated for Ontario that home wine production causes per capita wine consumption to be underestimated by 70 per cent and total alcohol consumption by about six per cent (Single and Giesbrecht, 1979). Just what the comparable figures would be in other provinces cannot be estimated. Nor are there good estimates for the impact of home beer and spirits production. However, the importance of the last two is probably minimal. Consumption of non-beverage alcohol by skid-row inebriates probably results in an underestimation of consumption of less than one per cent.

The quantity of alcohol not captured in official sales data for various reasons means that per capita consumption figures have their limitations. In general, per capita consumption figures from sales are most useful for large jurisdictions (without heavy tourism or home production) and for examining trends over time. They tell us little about who actually consumed the alcohol or what their characteristics were. They do have the advantage of consistency over time and place; most provinces and countries have per capita consumption figures that can be relied upon to be reasonably accurate within the limits suggested above.

Another way of studying drinking in Canada involves observational methods. These studies usually involve observing small groups of drinkers in taverns or bars, although some studies have been made at parties. Sometimes these observations are supplemented by interviewing individuals about their drinking. The advantage of such studies compared to other methods is their closeness to the drinking situation. They can tell us much about how people actually drink, how waiters and bartenders encourage drinking, what social interactions occur around drinking, and what the setting is like. They give a depth of detail and immediacy completely lacking in other methods. However, questions have been raised about how representative the observations are and to what extent they are affected by the interests or biases of the observer. Also, much drinking takes place in private and cannot be observed.

All methods of studying drinking have some advantages, and limitations. We should pay attention to data from all types of studies and take what each has to offer in the global picture. Fortunately, we have many representatives from each type of study in Canadian situations.

Surveys of Drinkers in the General Population

The Gallup Poll has been surveying drinking habits in the Canadian population for many years. The question asked in this survey is, "Do you ever have occasion to use alcoholic beverages, such as liquor, wine or beer, or are you a total abstainer?" Currently, about 73 per cent of adults report themselves to be drinkers, a large increase since 1943 when only 59 per cent were drinkers (see Table 4.2). However,

Table 4.2

The Proportion of Drinkers in the Population of Canada: 1943-1995

	Percentage Reporting Drinking			
Year	Overall	Ages 18 To 29	Ages 30 To 49	Aged 50 & Over
1995	73	77	78	65
1990	79	87	81	71
1985	82	90	86	70
1980	74	83	79	61
1975	76	84	79	61
1969	67	73	74	56
1958	65	70	68	57
1949	65	67	70	56
1943	59	59	64	53

Note: In 1958 and earlier years, adults were those 21 years and over. In 1959 and later, adults are those 18 years and over.
Source: The Gallup Poll, 1943-1995.

the proportion of drinkers, overall and in all age groups, was at its peak in 1985. Since that time declines have been seen in all age groups.

The Gallup survey, in common with many others, found fewer drinkers among those over 50 and especially over 60. The lower figure is partly explained by the poorer health of the older population; people with serious heart or circulatory problems are often advised to give up drinking. However, many older people grew up in the 1930s and 1940s when drinking was less common in Canada and temperance feelings were much stronger. A major increase in percentage of drinkers among those aged 18 to 29 came in the early 1970s (73 per cent in 1969 to 84 per cent in 1975). This likely occurred because many provinces reduced the drinking age from 21 to 18 between those years.

The *National Alcohol and Other Drugs Survey* (Eliany, Giesbrecht et al., 1992) is the largest survey of drug use in Canada and it gives far more information than do the earlier Gallup surveys. About 78 per cent of Canadians were drinkers in the NADS survey, although the question asked was different than in the Gallup studies. We might also note here that the NADS survey under-represented young males and failed to account for much of the alcohol sold in Canada.

If we set aside the problems of the NADS survey and take it at face value, it provides considerable information on drinking. As with all Canadian surveys, it showed that males were more often drinkers than females (83.8 vs. 71.8 per cent). Also, males drank more than twice as many drinks per week as females (5.3 vs. 2.0). They were also far more likely than females to drink five or more drinks on a single occasion (63.4 vs. 36.4 per cent). As with almost all Canadian drinking surveys, rates of drinking were highest in British Columbia, Ontario, Quebec and Alberta and lowest in Prince Edward Island, New Brunswick and Nova Scotia. This was true both for the proportions of current drinkers and the average number of drinks consumed. However, the NADS survey did not include the Yukon and Northwest Territories where alcohol consumption is usually higher than in any province. In the NADS survey levels of drinking were higher among those with higher educational levels, higher incomes, employed people and those who are single, separated or divorced. There were no major differences between those who spoke French and English.

Frequency of Drinking

Although most Canadians drink, they do so infrequently. In the NADS survey, only 10.7 per cent drank at least four times a week, and only 30 per cent drank more than once a week. In that survey, 38.2 per cent reported having one to seven drinks in the week prior to the survey, 7.5 per cent had eight to 13 drinks, and 7.5 per cent had 14 or more. Relatively few females drank as many as 14 drinks a week (about 2.6 per cent), but about 11.8 per cent of males did. The group with the highest proportion of heavy drinkers was males in the 20-24 year age range. Incidentally, those figures for consumption are much higher than in the *Canada Health Survey* done in 1979.

Data from a recent household survey in Ontario (see Table 4.3) showed that in 1991, the highest frequency of drinking was more common among certain segments of society (Adlaf, Smart et al., 1991). Almost three times as many males (10.2 per cent) as females (3.7 per cent) were daily drinkers. Daily drinking was most common among people aged 50 or older and lowest among those aged 18 to 29, however the number of drinking occasions showed much less variation across age groups. Daily drinking was heaviest in Toronto and the North and this correlated with the number of drinking occasions. Although there was no difference in daily drinking by education, those who had post-secondary education were most likely to be drinkers and to drink more frequently. However, those with higher incomes and in the professional groups were also more likely to be drinkers, and frequent drinkers than were people in the laborer, housewife or student categories.

Trends in Drinking

The Ontario surveys are one of the best ways of examining trends over time among Canadian drinkers. There is no consistent national survey of drinking using the same questions and methodology and detailed questions about patterns of drinking. However, the Ontario surveys began in 1977 and do use the same methodology.

The basic trend data for drinking in Ontario are shown in Table 4.4. Overall the proportion of drinkers did not change much between 1977 and 1991 (from 79.9 to 82.1 per cent). However, there was a very large decrease in the proportion reporting daily drinking (from 13.4 to 8.3 per cent). The proportion consuming five or more drinks on a single occasion fluctuated around 50 per cent with no real trend obvious, except that levels seemed to peak in 1982. It appears from the Ontario surveys that people became more cautious about their drinking in the late 1970s and have moderated their consumption since. What data we have from national surveys supports this view.

Per Capita Alcohol Consumption

It is a worldwide practice to report absolute alcohol consumption figures on a population basis. Usually, this is done in terms of litres of absolute alcohol (i.e., 100 per cent alcohol) consumed yearly on the average by each person aged 15 and over living in the area covered. It is assumed that this is the drinking population, although some people younger than 15 drink, and there are some abstainers among those older than 15. Of course, no one drinks 100 per cent alcohol, and it cannot even be obtained in beverage form. The figures are obtained in Canada by determining the amounts of beverages sold per person and assuming that beer is five per cent alcohol by volume, wine 13 per cent, and spirits 40 per cent. Of course, these amounts are a little arbitrary, as the actual concentrations of alcohol in different beers, wines, and spirits vary considerably. For example, regular-strength beers have between four and 5.6 per cent alcohol, while light beers may have from 2.4 to four per cent and near beers less than one per cent. "Strong beers" may contain up to 8.5 per cent alcohol. The alcohol content of wine varies from 6.5 to 14 per cent for table wines, but up to 20.3 per cent for fortified wines such as sherry or port. Spirits have a narrow range of alcohol concentrations from 39.5 to 40.5 per cent.

Table 4.5 shows the per capita alcohol consumption in Canada and the various provinces (Single, Williams and McKenzie, 1994). The average consumption in 1990-1 was 8.8 litres per person aged 15 and over. This equals about 513 drinks a year, or almost 10 per week, if we assume "drinks" to be 1 oz. of spirits, 12 oz. of beer, or 5 oz. of table wine. On the average, Canadian adults now drink less than two drinks per day. The variation across the country is considerable but about the same as found in the United States. In New Brunswick, Nova Scotia, P. E. I., Quebec and Saskatchewan per capita consumption is slightly lower than average, with New Brunswick being the lowest. Consumption in Newfoundland, Ontario and Manitoba is slightly above the average. Alberta, British Columbia, and the Northwest Territories are considerably above average (about 10 litres). Yukon is in a class by itself with a per capita alcohol consumption of 16 litres — nearly twice the national average and equal to about three and a half drinks per day for every adult. In general, in Canada the farther west and the farther north you go, the higher the level of alcohol consumption. In the Maritimes, especially Nova Scotia and P.E.I., some vestiges of temperance teachings still seem to remain.

Trends in per capita alcohol consumption were not very remarkable between 1974 and 1983. The rate of consumption was essentially

Table 4.3

Frequency of Alcohol Use During the Past 12 Months Among Ontario Adults - 1991

Characteristic	Prevalence (% using)	Never	<1 Per Month	1-3 Times Per Month	Once Per Week	2-5 Times Week	Daily	Times Per Week
Gender								
Male	88.2	11.8	12.6	24.6	18.8	21.9	10.2	1.80
Female	76.5	23.5	26.8	24.8	12.2	9.1	3.7	0.85
Age								
18 - 29 years	87.2	12.8	16.6	34.3	16.0	17.8	2.5	1.15
30 - 39 years	85.2	14.8	17.4	28.3	16.1	19.0	4.4	1.29
40 - 49 years	85.8	14.2	21.4	22.9	18.2	14.7	8.6	1.43
50 - 64 years	76.3	23.7	21.4	14.8	16.8	13.2	10.1	1.43
65+ years	73.8	26.2	28.5	16.2	7.3	8.6	13.2	1.40
Region								
Metropolitan Toronto	82.1	17.9	14.4	28.8	13.5	14.8	10.6	1.54
Metropolitan Outskirts	83.3	16.7	29.2	22.1	13.7	13.8	4.5	1.09
Eastern Ontario	83.1	16.9	19.3	20.6	18.0	19.9	5.2	1.37
Western Ontario	79.4	20.6	245	23.5	166	10.6	4.2	0.96
Northern Ontario	87.4	12.6	23.2	24.8	13.7	19.8	6.01	1.41

Education								
Elementary	71.6	28.4	30.5	16.6	10.4	6.9	7.2	0.97
Secondary	80.4	19.6	22.5	24.8	14.5	12.0	6.7	1.18
Post-Secondary	86.3	13.7	15.8	26.4	17.2	19.9	7.0	1.50
Occupation/Work Status								
Professional/Managerial	90.6	9.4	12.2	30.3	15.9	25.1	7.1	1.70
Sales/Clerical	82.0	18.0	20.5	28.1	13.7	14.5	5.1	1.16
Labor	82.3	17.7	18.2	20.2	21.0	15.5	7.5	1.38
Unemployed	85.5	14.5	29.9	10.6	22.0	13.0	9.9	1.47
Homemaker	68.0	32.0	31.7	17.6	9.5	6.2	3.0	0.64
Student	86.5	13.5	12.7	37.4	17.8	16.0	2.5	1.12
Other	80.8	19.2	25.6	17.6	13.7	9.0	14.8	1.59
Gross Family Income (in thousands unadjusted)								
<10	73.3	26.7	28.0	14.6	20.8	6.6	3.3	0.79
10 - 14.9	71.5	28.5	19.2	22.9	18.0	7.4	4.1	0.84
15 - 19.9	83.1	16.9	28.0	21.9	9.6	15.5	8.1	1.32
20 - 39.9	80.3	19.7	19.4	28.2	10.7	14.2	7.7	1.30
40 - 49.9	88.7	11.3	26.5	24.2	13.7	16.5	7.7	1.41
50+	89.1	10.9	16.5	24.4	18.4	21.9	7.9	1.64

Note: Underlined 1991 entries represent significant category differences in drug use.
Source: Adlaf, Smart and Canale, 1991.

Table 4.4

Trends in Drinking of Alcoholic Beverages in Ontario, 1977-1991

	Per Cent					
	1977	1982	1984	1987	1989	1991
Drinking in Past Year	79.9	77.6	84.5	83.1	82.6	82.1
Daily Drinkers	13.4	10.7	12.9	11.8	10.0	8.3
5 or more drinks (at least once past year)	52.5	57.8	49.2	54.5	54.6	50.1

Table 4.5

Per Capita Alcohol Consumption in Canada in Litres of Absolute Alcohol: 1974-75 to 1990-91 Ages 15 or Over

Province	1974-75	1978-79	1982-83	1990-91
Nfld.	10.45	10.88	10.60	9.20
P.E.I.	10.05	11.14	9.57	7.54
N.S.	10.00	10.40	9.85	8.12
N.B.	8.95	9.38	8.80	6.64
Que.	10.27	10.42	9.19	7.79
Ont.	11.45	11.52	11.16	9.3
Man.	11.36	10.74	10.85	8.00
Sask.	10.41	10.22	9.98	7.39
Alta.	11.82	12.79	12.94	9.44
B.C.	13.00	12.77	12.83	10.16
Yukon	21.14	19.87	19.39	16.00
N.W.T.	15.41	13.90	13.90	11.46
Canada	11.14	11.28	10.82	8.80

stable in the country as a whole. Virtually all provinces showed the same sort of stability, with less than a 5 per cent increase or decrease. However, remarkable changes occurred in the 1980s with a decline in consumption in every province and territory. Overall alcohol consumption declined in Canada by about 22 per cent between 1978 and 1990. The reasons for this decline are probably complex. They relate to changes in lifestyles promoting healthier eating and more attention to health in general (Smart, 1987). The decline in drinking also occurred at a time when people were smoking fewer cigarettes, exercising more and becoming more fit. It is interesting to note, however, that these changes occurred when prices for alcoholic beverages in real terms showed no decline (Smart and Mann, 1995). Perhaps the changes relate to real changes in attitudes about drinking and not merely reactions to economic factors or the recession, since prices were not related to alcohol consumption. Whether these declines in consumption will continue is, of course, a matter for speculation and further study.

The decline in alcohol consumption in Canada mirrors trends in a number of Western countries (Smart, 1989). The declines have been largest for spirits consumption and smallest for beer. There has also been a tendency for countries to homogenize their types of consumption, so that the proportions of beer, wine and spirits consumed become more equal in all countries. These changes probably relate to the stabilization in real income worldwide that has existed since about 1974, and perhaps to changes in lifestyles that place more emphasis on fitness, good nutrition, and staying healthy. In the past, increases in drinking have usually related to increases in affluence. For example, consumption in Canada and elsewhere rose greatly in the years 1950 to 1975, which were years of growing affluence and rapid increases in real income. Of course, the increase in affluence and real incomes stopped in Canada in the late 1970s. Unfortunately, the reasons for declines in alcohol consumption have not been much studied in other countries. For example, there are few studies of changes in prices of alcohol in other countries.

Expenditures on Alcohol

One reason for the increase in alcohol consumption between the 1950s and the 1970s was that alcoholic beverages became cheaper, at least in relative terms. It has been established that by the mid-1970s it cost far less in "real" dollars (that is, taking account of inflation) to buy alcoholic beverages than in the past. For example, in 1955 (the first year for which we have data) it cost 5.37 per cent of the average worker's disposable income to buy 10 litres of absolute alcohol. By 1979 it cost only 2.37 per cent of the average worker's income to buy the same amount. Average dollar expenditures on alcoholic beverages were about $350 per family for Canada in 1978; expenditures were lowest in the Maritimes and highest in Ontario, Quebec, Alberta, and British Columbia. Statistics for Yukon and the Northwest Territories were not available, but they were probably much higher not only because of the heavier consumption but also because costs for consumer goods are generally much greater in the North. However, since the 1970s the real cost of alcoholic beverages has not changed much, at least in Ontario, although people are now drinking less than a decade earlier (Smart and Mann, 1995).

The Distribution of Alcohol Consumption

Much research has been done to discover what proportion of people drink what amounts of alcoholic beverages — that is, the "distribution" of consumption. This research first started in France and was soon taken up by alcohol researchers in Ontario (Ledermann, 1956). The researchers have discovered that the shape of the distribution is very similar from one country to another and for many kinds of drinking groups. The distribution shown in Figure 4.1 illustrates the way populations usually drink. The graph was originally derived from a study of actual buying habits of a large group of Ontario residents in the 1960s so it is uncertain whether it is still valid (de Lint and Schmidt, 1968). The distribution is often termed "the single distribution," "the log normal distribution" or the Ledermann curve after Sully Ledermann, the French scientist who first discovered it and developed an equation for describing it.

There are a number of interesting aspects to this distribution. It is obvious that there are many users of small amounts, or light drinkers, fewer moderate drinkers, and even fewer heavy drinkers. Light, moderate, and heavy users are not sharply differentiated from each other; rather each category shades into the next such that the distribution curve is continuous and smooth. Ledermann and others have also found that there is a very close association between per capita consumption and the proportion of heavy drinkers. Suppose we take 10 cL of absolute alcohol (equal to 5-6 drinks) per day as the "heavy" level — a level that has been associated with some physical problems when maintained over a long period. If the per capita consumption is 8.8 litres — the current level in Canada — then about five per cent of the drinkers should be in the heavy category. However, if the level is

16 litres, as in Yukon, about 12 to 15 per cent should be heavy drinkers. As per capita consumption doubles, the proportion of heavy drinkers will be higher by a factor of 2.5 to 3.0. The implications of the distribution for prevention will be considered in more detail in later chapters. In general, it appears that the proportion of heavy drinkers cannot be decreased without an overall decrease in consumption.

Figure 4.1

Distribution of Alcohol Use in Ontario (1960s)

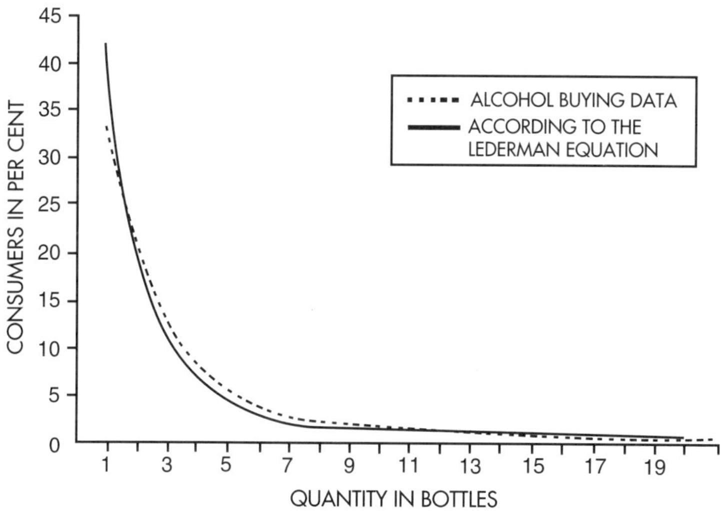

Beverage Preferences

In the country as a whole, beer is by far the preferred beverage, although the strength of the preference differs from place to place (Brown, Devan and Wallace, 1982; Brazeau and Burr, 1992). About 56 per cent of the alcohol drunk was beer in 1990, with 29 per cent spirits and only 16 per cent wine. The preference for beer is greatest in Quebec, Newfoundland and New Brunswick (59.5 to 63.7 per cent) and least in Alberta (49.1 per cent). Spirits are most popular in the Northwest Territories and Saskatchewan, and least popular in Quebec (only 15.8 per cent of alcohol sold). Wine is most popular in the wine-producing provinces of British Columbia, Ontario and Quebec, but is also popular in Yukon and Alberta. The popularity of beer has increased greatly in the past decade (49 to 56 per cent), while wine has had a remarkable increase (nine to 15 per cent). Spirits have shown a considerable decline.

Canadians also have shifting preferences for particular types of beverages. There is a preference now for lager (55 per cent of beer sales), which displaced ale. The preferred liquor is rye or "Canadian" by a wide margin (40 per cent of sales), followed by rum (16.4 per cent) and vodka (12.3 per cent). There is also a long-term trend away from fortified wines toward lighter table wines. Imported wines have more than half the market now, although they were insignificant before 1960.

Some International Comparisons

It is not easy to make international comparisons of alcohol consumption. Many countries do not conduct surveys, and some of those surveys that are done are difficult to interpret. Surveys in other countries use a variety of methods of sampling and different ways of asking about drinking. Hence, the results cannot be easily compared. For some countries, there are per capita alcohol consumption figures, but for many developing countries there is little interest in gathering accurate data on consumption. Also, per capita consumption figures are only for recorded production or sales reported to the government. In many countries, there are vast local industries producing illegal alcoholic beverages as well as many people producing small amounts for their own consumption. Pulque in Mexico and sugar-cane-based liquor in all South American countries are examples. Native beers and palm wine in Africa are also home-made and would not often get into government records. Since the official figures would greatly underreport alcohol consumption, comparisons with countries like Canada, that have minimal home and illegal production, might be misleading.

Probably the best comparisons of Canadian alcohol consumption are with the United States, Australia, New Zealand, and the European countries. These countries usually have accurate consumption data and minimal amounts of illegally produced or home-made alcohol. Data from such countries are shown in Table 4.6 (PGD, 1993). Canada's per capita consumption in 1991 was slightly lower than that in the United States, but substantially higher than that in Scandinavian, African and most South American countries. However, Canadian consumption is much lower than that in the wine-producing countries of France, Italy, Spain, and Switzerland. We have consumption data for 1970 to 1990 for only 15 Western countries: 11 European countries, the United States, Australia, and New Zealand as well as

Table 4.6

Per Capita Alcohol Consumption in Various Countries: 1970, 1977 and 1990

	1970	1977	1990
SOME WESTERN COUNTRIES			
Canada	6.45	8.53	7.1
United States	6.84	8.13	7.4
United Kingdom	5.23	6.75	7.6
Belgium	8.94	10.06	9.9
Czechoslovakia	9.12	9.91	8.8
Federal Republic of Germany	11.23	12.20	10.6
Italy	14.45	12.43	9.0
Spain	11.26	12.81	10.8
Switzerland	10.52	10.57	10.8
France	19.59	17.31	12.6
Luxembourg	10.19	14.37	12.2
Norway	3.57	4.45	4.1
Sweden	5.64	6.04	5.5
Australia	8.22	9.81	8.0
New Zealand	6.66	8.36	7.8
SOME SOUTH AMERICAN COUNTRIES			
Mexico	2.14	2.36	3.0
Argentina	13.08	14.02	7.5
Chile	6.51	7.08	6.4
Venezuela	3.68	4.06	3.3
SOME ASIAN AND AFRICAN COUNTRIES			
Japan	4.94	5.46	6.5
Malaysia	6.64	6.29	.50
Turkey	.54	.86	.70
Tunisia	.76	.51	.50
South Africa	4.30	5.16	4.9
Morocco	.36	.29	.30
Algeria	.33	.27	.30

The figures are based on total population, not population 15 and over as in Table 4.3.
Conversion figures were beer - 5%, wine - 12%, vermouth etc. - 18%, spirits 40%.
Sources: Addiction Research Foundation, 1984; Produktschap voor Gedistilleerde Dranken, 1993.

Canada. In 1970 and 1977 Canada ranked in the middle of the group, but by 1990 was very close to the bottom of the list. Declines in alcohol consumption through the 1980s have been greater in Canada than in most other countries.

If we extend the comparison of Canada to non-western countries, we have many problems, in that most African and Asian countries have no data or only questionable data. We will therefore make such comparisons tentatively. What we have suggested is that Canada has a higher alcohol consumption than Asian countries, higher than all South American countries except Argentina, and higher than all African countries for which we have data. Here we should also remember that most countries do not have data that are consistent over time. On a worldwide basis, Canada would seem to rank about 13th in per capita alcohol consumption out of the 26 countries for which trend data are available. If we consider all 48 countries for which 1990 data are available, Canada falls in the middle at number 24. Of course, many developing countries do not report or underreport their true alcohol consumption, and Canada's real ranking may be much lower on a worldwide basis.

Observational Studies

Not many studies have been made of Canadian drinkers as they actually do their drinking. Most Canadians do most of their drinking at home or at the homes of friends — about 70 per cent of alcohol sold is for "off-premise" consumption. However, all of our observational studies have been carried out in licensed establishments such as bars, taverns, and beer parlors. Unfortunately, we have no published studies of drinking in family or party situations. Nevertheless, observational studies do give insights into drinking that cannot be obtained from government statistics, providing a depth of understanding and a feeling for socio-cultural factors.

One early study of beer drinkers in Edmonton beer parlors, for example, showed that isolated drinkers drank less than those in groups and stayed for shorter periods of time (Sommer, 1965). However, "loners" who were eventually joined by others stayed the longest. Group drinkers drank more than loners and they stayed for longer periods of time. In general, the amount people drank depended upon the they spent there, not on whether they were alone or not. Sommer argued that people go to pubs not to order a specific number of drinks but to meet other people. When they do, they stay longer and drink more in the process. The main reasons for going out to drink seem to be social; if social needs are not fulfilled, the drinking seems to stop and patrons leave.

Observational studies of patrons in Vancouver beer parlors confirm Sommer's findings (Storm and Cutler, 1981). Researchers found that most patrons drank in groups of three to five persons. Men predominated. Men were aged on average about 34 and women about 29. Isolated drinkers tended to be older men drinking for short periods in the morning or afternoon hours. Most drinkers had two beers or less, but about 18 per cent had five or more. About 14 per cent were estimated to have blood alcohol concentrations of 80 mg/100 mL (the legal limit for driving) or more when they left. On average, patrons drank one drink about every 23 minutes. Their total consumption related closely to how long they stayed. The rate of drinking was relatively constant, although there was some tendency for the first drinks to be taken more quickly.

Ratcliffe and Caverhill (1978) also studied drinkers in Edmonton bars, taverns, and pubs. They found that people in manual laboring jobs constituted the largest proportion of patrons in "beverage rooms" serving beer only. Young adult or college groups and "young toughs" were also prominent, but office workers were less numerous. The remainder were elderly persons and native people.

Ratcliffe and his colleagues made several studies of how much people drink in various places. Games and dancing contributed to drinking only if patrons participated, as the tendency was for participants to stay longer and drink more. Those in "stand-up" bars were usually men who drank small amounts and left quickly. Ratcliffe also studied drinking in comfortable "lounges" and the more spartan "taverns." He found that the more casual and comfortable lounges did not have lower drinking levels: patrons in lounges and in taverns drank at the same rate. Female tavern patrons stayed longer and drank more than female lounge patrons.

Ratcliffe also made a follow-up study of the same beverage rooms studied by Sommer in 1963 (Ratcliffe, Nutter et al., 1980, 1982). After 15 years the beverage rooms had been modernized and the rules changed to allow women patrons, as well as some games, although no one played them. Also, a wider variety of beverages were allowed; in Sommer's time only draft beer could be sold. Ratcliffe found that patrons drank much more slowly in 1978 than in 1963. Although they stayed about the same length of time, they drank less overall. It is difficult to be sure why patrons drank more slowly in 1978, but a variety of amenities had been added and the price had increased remarkably.

The presence of women also probably had a moderating effect, as several studies have shown that "men-only" drinking is much heavier than mixed drinking.

Popham (1982) studied patrons in a variety of bars and taverns in Toronto in the 1960s. He found that most taverns had a regular clientele plus some infrequent users. Regulars were usually men, married, in professional or white-collar/sales jobs. They usually came to the tavern more than once a week. Their typical consumption was four glasses of draft beer, one every half hour, for a total intake of 32 oz. A few non-regulars and female customers drank at about half that rate. Drinking speeds were faster in "skid-row" taverns. The rate of drinking was governed by the waiters, who brought more drinks according to a schedule they, not the drinker, established. Most regulars went to taverns to escape from loneliness, make and meet friends, or relax after a tiring day, although some were alcoholics and went primarily to drink.

Elliott and Cosper (1982) have made studies of drinking in Halifax taverns and pubs that confirm some of Popham's earlier observations about "regulars." In a study of records kept by participants, they found that only about six per cent of respondents went to a pub or tavern on the average day. About 39 per cent of respondents in a survey said they never went to pubs or taverns, and an additional 27 per cent went only a few times a year. Only 15 per cent went as often as once a week and could be considered "regulars." However, they made up a large proportion of tavern patrons and accounted for more than 60 per cent of the total visits. About one-third of the people in taverns at any particular time seemed to be those who went on a daily basis.

Some attention has been paid to heavy drinking occasions or "festival" drinking in Canada. It must be obvious to any observer that much drinking takes place at Christmas and New Year's, and sales figures bear this out. Also, much heavy drinking occurs at weddings, graduations and other special events. Minor deviance and drunkenness are often allowed by "polite society," and even the police will suspend the usual arrest procedures for special festivals. Listiak (1974) studied drinking in Hamilton pubs during Grey Cup celebrations and found that the "middle classes" showed more deviance than the "lower classes" — that is, more heavy drinking, aggressiveness, and "minor immorality." He hypothesized that festival-related moral holidays are more functional for the middle classes, who require more "release from everyday cares." Ossenberg (1969) reached a similar conclusion in his study of Calgary taverns during the annual Calgary

Stampede. Middle-class customers were more inebriated, more raucous, and more sexually expressive than were the lower classes. He suggested that Stampede Week excesses are "more functional for people who tend to be inhibited in their daily lives."

While there have been studies of tavern drinking, other places for drinking have not been much studied. Recently, more attention has been paid to how the place of alcohol consumption relates to drinking in Canada. Single and Wortley (1993) used the NADS data to look at all types of drinking. Most drinking (75 per cent) was done at home. The average number of drinks per occasion was very low (2 drinks) for drinking at home, but higher for drinking in bars or taverns or at parties (more than 3 drinks). Males and young people drank more often outside the home, usually in bars or taverns. Also, heavier drinkers and those with problems were more likely to drink in bars or taverns. However, it was not clear whether heavier drinkers were drawn to bars and taverns or whether the setting actually made people drink more. More study is needed of drinking at party situations such as weddings and special events, even though these do not make up a large part of most people's drinking.

Summary

Most people in Canada drink alcoholic beverages but prefer small amounts. There are numerous methods of studying drinking, and all have been used in Canada. These include use of government data on per capita consumption, surveys of the general population, and observational studies of people in bars and taverns. All methods have advantages and disadvantages. Currently, surveys show that about 73 per cent of Canadians are drinkers, compared to about 59 per cent in 1943. Only 10.7 per cent drink four or more times a week, and 30.4 per cent drink more than once a week. Per capita annual consumption is 8.8 litres, or about 10 drinks a week on the average. Drinking rates are generally higher in the North and the West. Typically, Canadians prefer beer, followed by spirits and wine. At present, Canada's alcohol consumption ranks about 13th among the 26 countries in the world with trend data. However, if we include developing countries where less alcohol is usually consumed, Canada would be nearer the top.

Observational studies are not very helpful in determining quantities consumed, as about 70 per cent of consumption occurs at home. However, some have yielded interesting information. Drinkers in groups drink more than loners, chiefly because of the length of their stay. Drinking rates in a tavern tend to be constant, with one drink taken every half hour or so. The presence of women seems to have a moderating effect on drinking behavior. Studies of heavy drinking occasions, such as Grey Cup celebrations, show that more "middle-class" than "lower-class" people are involved. Males, young people and heavy drinkers drink more often in bars and taverns as well as parties; we are not sure whether the context makes them drink or whether certain heavy drinkers are more drawn to particular situations.

Chapter 5

Drinking Among Specific Groups of Canadians

In the last chapter we looked at alcohol use in the Canadian population as a whole and noted how drinking habits vary by region and by some social groupings. In this chapter we will look more closely at the drinking habits of women, youth and other groups that have been studied by Canadian alcohol researchers. We will also consider the characteristics of Canadians who drink very heavily to see what, if anything, they have in common.

Gender-related Differences in Drinking Patterns

Drinking norms for men and women have traditionally been quite different. Women have usually favored moderation or abstinence and played important roles in the temperance and Prohibition movements. Both women and men tend to disapprove of women who drink heavily and get drunk but men are more tolerant of male heavy drinking and drunkenness (Ferris and Templeton, 1993). The general tendency of women to be more disapproving of heavy drinking than men is likely due, in part, to the fact that many women have been victimized by drunken fathers and partners.

Although Canadian women's drinking habits have changed significantly in the past few decades, we saw in Chapter 4 that women are still less likely to drink alcohol than men and that fewer women are heavy drinkers. Most indicators of alcohol problems also suggest that there are fewer female than male problem drinkers. For example, twice as many Canadian men as women die from liver cirrhosis, and four times as many die from alcoholism. Women are also much less likely to be admitted to hospital suffering from cirrhosis and alcoholic psychosis (Single, Williams and McKenzie, 1994) and the ratio of males to females in treatment services ranges from two to one to four to one, depending on the type of service (Ellis and Rush, 1993).

Direct comparisons between of the drinking habits of men and women and their consequences may, however, be misleading. For example, women do a greater proportion of their drinking at home and are thus less likely to experience some types of alcohol problems (drinking and driving, work-related problems or public intoxication). However, women are more vulnerable to the physical effects of alcohol due to lower body weight and a higher ratio of body fat to water than men (Lex, 1991). As a result, women typically need to drink less than men to experience comparable effects.

Differences between men and women who experience drinking problems have been studied in general population surveys and surveys of people in alcoholism treatment programs. Surveys show that both men and women who experience drinking problems tend to be young, single, and not religiously observant (Smart and Liban, 1981). However, women are more likely than men to report drinking alone and less likely to report drinking in bars. Neither men nor women are likely to seek help from specialized addiction programs, but women seem to make even less use of these programs than men (Rush and Tyas, in press). This may be because treatment programs do not always cater to women's needs and situations (Allen, 1994). On the other hand, it may be that women with alcohol problems are more likely than men to seek help from doctors for "physical" or "mental health" problems in which alcohol is a factor, rather than seeking treatment for alcohol problems per se.

The reports of patients in alcohol treatment programs show that women start drinking later than men and develop drinking problems more rapidly (Smart, 1979b; Ashley, Olin et al., 1977). Women's drinking problems also seem to be more likely than men's to develop in response to personal crises (e.g., divorce or death in the family). Male problem drinkers, on the other hand, are more likely to report that they always drank a great deal and experienced alcohol problems from an early age. Such differences have led some experts to describe women's drinking as more "reactive" than men's. This is consistent with the results of a Toronto study that showed men in an addiction treatment program were more likely to be diagnosed as having a personality disorder than were women. In contrast, women were more likely to be diagnosed as suffering from anxiety, psychosexual disorders or bulimia (Ross, Glaser and Stiasny, 1988).

Women seeking help for drinking problems describe problems in their backgrounds similar to those of men (parental alcoholism, divorce, physical abuse and psychiatric illness). U.S. studies of women in treatment reported high rates of sexual abuse (Wilsnak, Wilsnak and Klassen, 1986). Canadian clinicians and researchers also report that many Canadian women with alcohol problems have been abused (Aylward, 1994).

There is no good evidence to suggest that problem drinking among women is related to "sex-role confusion" (i.e., not knowing what society expects) or "role conflict" (playing many different and potentially conflicting parts in the system). Women with drinking problems cannot be distinguished from those who do not experience problems based on their views on the role of men and women (see Jung, 1994 for a recent summary of research on this topic).

Age-related Differences in Drinking

The legal drinking age is 18 in Quebec, Manitoba, Alberta and P.E.I. and 19 elsewhere. However, drinking below the legal drinking age is quite common. We saw in Chapter 4 how drinking generally varies by age and in this chapter we will look more closely at the drinking habits of two, widely separated age groups; youth and the elderly.

Drinking among Young People

In the 1970s concern about young people's drinking became widespread, although it had been identified as an issue by teachers and parents since the 1950s. In the early 1970s most provinces lowered the legal drinking age from 21 to 18 or 19, a change that was followed by increases in young people's drinking and drinking problems (Smart, 1993). The lowering of the legal drinking age also sparked concerns that young people may be especially vulnerable to the toxic effects of alcohol and might develop inappropriate drinking habits that would be difficult to break.

There have been many studies of drinking among young Canadians and some are mentioned later in this chapter and elsewhere in this book. A detailed review of these studies has also been written by the first author (Smart, 1993). Here we will consider what these studies have told us about where and why young Canadians drink and what they have had to say about drinking. However, it is also important to note that since the late-1970s, surveys of high school students have reported a steady drop in the proportion of students who drink (Adlaf, Ivis et al., 1995). On average, students are also drinking less and having fewer alcohol-related problems than a decade ago. Heavy drinking, drinking and driving, and alcohol-problems among a substantial minority of young people are nonetheless a cause for concern. Reckless drinking episodes involving young people who are otherwise moderate drinkers are also worrying.

Studies in Ontario show that young people do much of their drinking in bars or taverns where they consumed an average of four drinks per occasion (Smart, 1993). More than half said that they drink at parties, social gatherings or weddings, occasions where they also had an average of four drinks. Far fewer young people reported drinking during a quiet evening at home, while visiting friends, having friends in, or at meetings, clubs or sporting events. When drinking occurred in these situations, it typically involved fewer than two drinks.

For 15 to 19 year olds, who cannot drink legally, more than half of their drinking occasions are in bars, taverns or at parties. These situations, where parental monitoring is low and heavy drinking is likely, are an important target for prevention programs.

Many adolescents drink only on special occasions such as holidays, weddings or other family events. For example, about 20 per cent of students in one Ontario study said that they drank most at such events (Smart, Adlaf and Walsh, 1991). The tendency is for younger students to do most of their drinking at special family occasions; as they grow older, they do more of it away from such events and away from home. Students who drink heavily and have problems often drink in cars and parks (Smart, Gray and Bennett, 1978).

There are several ways of answering questions about why young people drink. We could look at factors that are statistically correlated with drinking and try to make some sense of them. We could look at the situations in which drinking occurs or at the behavioral and emotional changes that occur following drinking, or we could simply ask young people why they drink. This is perhaps the most valid approach and several Canadian studies have included simple, direct questions about reasons for drinking.

Answers to these questions indicate that most young people first drink to find out what it is like and to be sociable and that many had their first drink with parents. Few report drinking because of social pressures and most drink only in social settings. However, in some studies up to 20 per cent of young people say that they to drink to deal with problems such as shyness, nervousness, tension or to forget their worries.

Most young Canadians disapprove of heavy drinking. One study found that 95 per cent of young people agreed that it was dangerous to drink and drive (Garceau, 1985). A large majority also agreed that alcohol affects reaction time, is harmful to the unborn, makes it hard to concentrate, is habit forming and slows you down. However, more than 80 per cent of those interviewed drank and 39 per cent drank every week. About one in four high school students reports driving after drinking (Adlaf, Ivis et al., 1995). Another study

(Siggner, 1988) indicated that few young people believe that moderate drinking can be good for your health but 32 per cent said they did not mind getting intoxicated once in a while.

The most recent national study (Eliany, Giesbrecht et al., 1992) showed that most young people do not believe that you should drink enough to feel alcohol's effects in most situations, including parties, sports events, drinking with friends or co-workers at lunch, after work, or even having drinks at home. Thus, Canadian youth are more conservative in their attitudes than in their behaviors. If their drinking reflected their attitudes, they would drink less and have fewer problems.

Two groups of young people that have attracted special attention from alcohol researchers are "street youth" and university and college students. Both are popularly believed to include many heavy drinkers, a view that research tends to support. Thus, in three different studies six and nine per cent of street youth were found to be daily drinkers compared with one to two per cent of people of the same age in the general population (Smart, Adlaf et al., 1990; 1992; Radford, King and Warren, 1989).

Drinking among University Students

Two large-scale studies of Ontario university students have been reported by Louis Gliksman and his colleagues at the Addiction Research Foundation. The most recent study was conducted in 1993 and involved over 6,000 students from six universities (Gliksman, Adlaf et al., 1994). The results showed that almost 95 per cent of students drank alcohol. Most (69 per cent) reported drinking less than 15 drinks a week. However, about 20 per cent of males and five per cent of females reported drinking over 28 drinks a week. Also the majority of male and female drinkers reported that, at some time during the previous year, they had a day of heavy drinking (five or more drinks).

Among both males and females heavy drinking was more common for first-year students, those living in university residences, those in arts programs and those with lower grades. Fourth-year students living at home, those in science programs and those with "A" grades had the lowest average consumption levels.

About 50 per cent of male and female drinkers had experienced hangovers in the two months prior to the survey and many had experienced other problems related to alcohol in the same period. The most common problems were nausea/vomiting (28 per cent of males, 22 per cent females); missing classes (20 per cent of males, 15 per cent of females); regretting behaviors while drinking (22 per cent of males,

20 per cent of females). Overall, these results were similar to those from a similar survey conducted six years earlier (Gliksman, Engs and Smythe, 1989).

Drinking Among the Elderly

We have already noted lower levels of drinking among older Canadians. This may be due to generational factors, premature deaths among heavy drinkers, or decisions to quit or cut down with advancing years. However, the long-term follow-up studies needed to show the relative importance of these and other factors have not been done. Indeed, very few Canadian studies have focused on drinking among the elderly, although there have been several such studies in the United States.

One Canadian household survey comparing elderly problem drinkers with elderly non-problem drinkers found most problem drinkers to be men. Foreign-born men were disproportionately represented among those experiencing alcohol-related problems (Adlaf and Smart, 1995).

Reasons given for drinking in Canadian national alcohol and drug surveys are generally similar for the elderly and other groups. The majority of drinkers of all ages say that they drink to be sociable, with meals or to relax and feel good. However, Canadians over the age of 65 are less likely to report "instrumental" reasons for drinking (to forget worries or to feel less shy) than those under 65. This is especially true for older women (Graham, Carver and Brett, 1992).

American studies of drinking among the elderly have raised more questions than they have answered and it is clear that there are many difficulties in studying drinking and drinking problems in older age groups (Graham, 1986). For example, memory loss and problem denial are more common among the elderly. Also some indicators of alcohol-related problems such as problems at work, drinking and driving offences, and public intoxication are not frequent among the elderly. Finally, in studies of drinking among the elderly, health problems associated with normal aging may be mistaken for alcohol-related problems.

It is certain, however, that some elderly people drink heavily and have drinking problems. Some of these are "late-onset" drinkers who turn to alcohol in response to the stresses of aging and other later life events, while others are long-term heavy drinkers. Heavy drinking among the elderly is of concern given the aging of the overall population and it is likely that demands for addiction programs for elderly will increase. The prospects of increasing numbers of elderly drinkers is also of concern because older people seem to be especially vulnerable to the adverse effects of alcohol. This is the result of factors associated with aging such as reduced body weight, general ill health and the frequent use of prescription drugs (Williams, 1984).

Drinking and Marital Status

Married Canadians and those who are cohabiting are slightly less likely to drink or to drink heaviliy than those who are single, separated or divorced (see Table 5.1). This is partly because there is a positive relationship between age and marital status although some studies show that both age and marital status have independent effects on drinking, and that young unmarried males are especially likely to drink heavily. Of course, divorce and separation may also lead to increased alcohol consumption and alcohol-related problems and, as noted earlier, women with alcohol problems often report starting to drink heavily in response to such crises. Canadians who are widowed are far less likely to drink or to drink heavily than those who are single, married or divorced. This is probably many widowed people are older females who generally drink less than others.

Drinking and Employment

Relationships between drinking and employment reported in the 1993 Canadian survey on alcohol and drugs (Single, Brewster et al., 1994) are summarized in Table 5.1. The data show that professionals and skilled workers are more likely to be current drinkers than other groups, while homemakers and retired people are the least likely to drink. Groups with the highest average levels of consumption are the unskilled and those looking for work. These groups also report more heavy drinking days.

Only American studies give more detailed information about drinking levels and drinking problems in specific occupational groups. The most recent U.S. study was conducted in 1988, involving almost 27,000 people (Parker and Harford, 1992). The results showed that white-collar occupations had the highest percentage of current drinkers. This was particularly true of senior managers and administrators, professionals, technicians and sales representatives. Although there were proportionately fewer current drinkers in blue-collar jobs they tended to have higher average daily consumption levels, especially those in the food service, health, construction, cleaning, and personal service jobs. Machine operators and laborers also tended to drink heavily. Among food service workers, bar tenders and waiters

Table 5.1
Drinking Status, and Drinking Patterns for Selected Groups[1]

	DRINKING STATUS		LEVEL AND HEAVY DRINKING OCCASIONS, AMONG CURRENT DRINKERS		
	Never	Current[2] Drinkers	Drinks/ Week[3]	% Any Heavy Occasions[4]	Mean # of heavy Occasions[5]
Marital Status:					
Single	8.3	78.0	5.0	63.4	18.6
Married or Cohabiting	6.8	75.1	3.9	41.0	13.4
Divorced or Separated	5.9	76.1	4.6	43.5	18.1
Widowed	15.4	48.3	2.7	14.9	6.8
Employment:					
Professional	3.8	87.5	3.5	40.1	9.5
Semi-Professional	3.5	84.7	4.2	49.9	14.2
Supervisor	5.0*	80.4	4.9	53.5	16.4
Skilled/Farmer	3.7	86.3	4.6	56.1	15.2
Semi-Skilled	6.3	78.2	4.7	54.9	18.1
Unskilled	5.8	79.2	5.0	54.6	17.4
Other Working	4.8*	75.8	3.9	44.0	21.4
Looking for Work	7.0	75.2	5.9	54.2	22.6
Student	11.7	71.9	3.8	63.1	13.4
Homemaker	14.4	59.8	2.2	22.8	10.3
Retired	8.9	60.0	4.2	17.8	16.4
Other	8.4	58.7	8.1	40.6	56.4
Community Size:					
>1,000,000	8.8	72.4	4.3	43.0	15.1
>500,000 & <1,000,000	6.0	79.2	4.1	46.8	17.5
>100,000 & <500,000	7.3	75.0	4.2	47.0	13.5
<100,000	6.5	75.7	4.2	46.4	15.2
Rural	8.3	72.3	4.2	49.7	16.8
Religious Attendance:					
Weekly	14.8	58.2	2.6	25.8	10.2
Monthly	6.8	74.7	2.9	36.9	11.5
Less than Monthly	5.1	81.8	4.1	52.2	12.9
Never	5.2	79.8	5.1	52.3	19.9
Not Applicable	5.2	79.7	5.8	58.3	18.7
Language:					
English only	5.2	77.7	4.2	47.8	15.8
French only	11.6	72.8	4.5	46.2	17.3
Other only	19.3	50.0	2.5	29.3	6.7
More than one	10.7	66.9	4.1	39.7	12.5

1 Source: Single, Brewster et al., 1994.
2 Respondents who have consumed at least one drink of alcohol in the past 12 months.
3 Mean number of drinks per week for current drinkers only.
4 Percentage of current drinkers reporting consumption of five or more drinks on at least one occasion in the past year.
5 Mean number of occasions in the past year on which five or more drinks were consumed, among those reporting at least one occasion.

had the highest average consumption levels. Among male white-collar workers writers, artists, entertainers and those in sales had the highest consumption levels. For women, the highest levels of drinking were found among computer scientists and mathematicians.

Symptoms of alcohol dependence were more common among adults in certain blue-collar occupations including food service workers, farm and forestry workers, construction workers, machine operators and laborers. Male bartenders were far and away the most likely group to be classified as alcohol dependent: of 36 bartenders interviewed, 37 per cent were considered to be dependent an alcohol while another 15 per cent were classified as "severely alcohol dependent."

Earlier studies conducted in Britain and North America (reviewed by Whitehead and Simpkins, 1983) suggest that a variety of factors influence rates of drinking and problem drinking in different occupational groups. These include: (1) no contact between the sexes during work hours; (2) workmates who approve of drinking on the job; (3) the likelihood of heavy drinkers being recruited to the workforce; (4) a preponderance of young workers; (5) the availability of inexpensive alcohol (free or subsidized); (6) workmates who approve of heavy drinking on the job; (7) management that tolerates heavy drinking on the job; and (8) social pressures to drink on the job. Work-related stress has also been cited for heavy drinking among police and physicians (*MacLean's* magazine, March 1994; Willenbring, Johnson and Tan, 1994).

Drinking in Communities of Different Size

Large cities have always been regarded as dens of iniquity, and city life is blamed for almost all social ills. In particular, widespread heavy drinking and drunkenness are often regarded as mainly urban problems. However, Canadians living in rural areas report quite similar levels and patterns of drinking to those in large cities. Moreover, there is no consistent relationship between community size and the percentage of drinkers, the average weekly consumption and number of heavy drinking occasions (see Table 5.1).

Drinking among People with Physical Disabilities

Studies of drinking patterns and problems among people with physical disabilities have produced conflicting results. Some studies suggest that people with disabilities drink more and have more problems than the general population (Hepner, Kirschbaum and Landes, 1980; Anderson, 1980-81) but other do not (Dean, Fox and Jensen, 1985; Meyers,

Branch and Lederman, 1988). However, it is difficult to compare the results of different studies because they have used different samples, definitions of disability and measures of alcohol use and problems.

The only large-scale Canadian study concerning substance use and physical disabilities was conducted in Ontario in 1989 (Adlaf, Smart and Walsh, 1992). It involved a survey of 869 adults over the age of 18. Of these, 31 (four per cent) reported that physical disabilities prevented them from working and 70 (eight per cent) reported that physical disabilities limited their participation in the work place. The remainder (88 per cent of the sample) reported no physical disabilities.

The study found a significant association between work limitations and substance use. Those reporting that they were unable to work drank less often than both those without limitations and those with disabilities that limited participation in the work force (labelled "moderate limitations" by the authors). Thirty-nine per cent of those with moderate limitations reported at least one alcohol problem, compared with seven per cent of those with total limitations and 24 per cent of those with no limitations. The high rate of alcohol problems among those with moderate limitations still held after controlling for age, gender and frequency of drinking. Among those with disabilities, drinking or drinking problems were not associated with the type or duration of limitation, or the need for assistance.

Although the study could not fully account for the high rate of alcohol problems among those with moderate disabilities, the authors speculated that they may experience low self-esteem. Unlike those with no limitations or those with total limitations, those with moderate limitations experience the tribulations of both having a physical limitation and being employed. This combination of work and disability may predispose them to depression and low self-esteem, factors that are strongly associated with alcohol misuse in their own right.

Drinking in Different Ethno-cultural Groups

Many Canadians identify themselves as members of particular ethno-cultural groups. These groups are to some extent distinguished by their histories and their current beliefs, values, customs and organizations. This subjective sense of membership in a particular cultural group is sometimes termed "ethnicity" (Cheung, 1990-91). The term is sometimes used interchangeably with "race," but this is quite wrong. The concept of race has little scientific value as it only concerns superficial differences (such as skin color) between people whose ancestors came from different parts of the world. However, these differences are not correlated with more profound biological differences, nor do they account for differences in behavior. Few academic researchers believe otherwise. The values, attitudes, behaviors, achievements and problems that characterize some ethnic groups have nothing to do with "race." Rather, they reflect the historical and current social conditions experienced by their members. It is also wrong to assume that superficial "racial" differences indicate identification with particular cultures. For example, black classical musicians sometimes find it amusing that they are expected to play jazz or other "black" music when they are more comfortable playing Bach.

While members of some ethno-cultural groups tend to have similar drinking customs, drinking has been studied in detail only among Canada's native peoples (see Chapter 6). However, a few studies have considered drinking and drinking problems in other groups. For example, one Vancouver study found that Chinese Canadians were significantly under-represented among people charged with drinking and driving offences (Lin and Lin, 1982). Although Vancouver had 50,000 Chinese Canadians at the time of the study, Chinese Canadians were only involved two alcohol-related road accidents. This reflects the fact the Chinese Canadians have tended to maintain the drinking customs of their country of origin, characterized by high rates of abstinence and drinking only at meal times. Drunkenness is also discouraged and punished by ostracism in traditional Chinese cultures.

However, concerns about alcohol have begun to emerge in Vancouver's Chinese community. This was evident from a needs assessment study reported in 1990. The study involved interviews with selected informants from Vancouver's Chinese, Indo-Pakistani and Latin-American communities (Legge and Sherlock, 1990-91). About 40 per cent of the Chinese informants thought that alcohol problems were "somewhat" or "very" widespread in their community and many indicated discrepancies between "acceptable" and "actual" drinking levels especially for youth and women. Concerns about drinking and about discrepancies between ideal and actual drinking levels were even more common among Indo-Pakistan and Latin American informants. Family difficulties were seen both as causes and consequences of alcohol problems in all three communities.

Another study, conducted in Toronto, showed that during the 10-year period ending in 1974 only 0.5 per cent of the alcoholic patients of the Addiction Research Foundation were Jewish compared with six per cent of the Toronto population during the same period (Schmidt and Popham, 1976). This is consistent with the results of

American studies and with everyday observations of low rates of heavy drinking among Jews. Jews are not required to refrain from alcohol for religious reasons and alcohol is used in Jewish ceremonies and family meals. In fact, very orthodox Jews have fixed patterns of alcohol use associated with religious customs. However, excessive drinking and drunkenness are strongly discouraged in traditional Jewish homes and institutions.

Other Canadian studies have focused on ethno-cultural differences and drinking among high school students. In one such study, Ontario students were asked in a written survey to indicate with which, if any, ethnic groups they identified, how often they drank and what alcohol-related problems they had experienced (Smart, Goodstadt et al., 1984). Those who identified with Eastern European groups (Czechoslovakian, Polish) reported drinking most often, while these who identified with South and East Asian groups reported drinking least often. Those reporting the highest levels of drinking and the most alcohol-related problems identified themselves as German or French. Those who drank least per occasion and reported fewest problems identified themselves as Italian or Asian.

In a second study involving Canadian high school students, ethnicity was assessed from the answer to a question about the ancestry of students' fathers and the language spoken at home (English or other). The results were similar to the first student study. Controlling for age, gender, language and place of residence, the frequency of drinking was higher among students whose fathers were described as East or West European and lowest among those whose fathers were described as black, Oriental or East/West Indian (Adlaf, Smart and Tan, 1989).

Language spoken at home distinguishes some ethno-cultural groups and the Canada-wide *1993 General Social Survey* showed that Canadians who do not speak either English or French at home are less likely to drink and less likely to drink heavily than those who speak either of Canada's two official languages (see Table 5.1). This finding reflects the fact that many of those who do not speak French or English at home are recent immigrants from countries where alcohol is forbidden or used infrequently.

Reasons for using alcohol are often quite distinctive in certain ethnic groups, and these reasons also influence levels of drinking and problem drinking. For example, wine plays an important role in Jewish religious observance, but here it is used primarily for its symbolic significance and not as an intoxicant. This may account the low prevalence of heavy drinking and alcohol problems among religious Jews.

Similarly, drinking in traditional Italian and Chinese cultures is very much tied to family meal times. Food reduces the chances of drunkenness because it delays the absorption of alcohol into the bloodstream. Also, associating alcohol with meals reduces the chances of excessive drinking, solitary drinking, and the linking of alcohol use with male competitiveness and violence. In contrast, Irish, Finnish, and some Eastern European cultural groups value convivial "time-out" drinking for men, contributing to more frequent public drunkenness.

The drinking attitudes and practices of cultural groups do not, of course, exist in a vacuum. Also important are the relationships between a group and the rest of society. Well-established and respected groups may not have the same cautious attitudes to alcohol use as insecure groups reaching for acceptance. On the other hand, members of groups of low social status, for whom things are bad or getting worse, may have few reasons for not drinking or getting drunk and may use alcohol to escape from their problems. New immigrants may be vulnerable to heavy drinking to cope with culture shock and a loss of identity. Those with little previous experience with alcohol may also be vulnerable to heavy drinking. However, these issues have not been researched in Canada.

More research is needed on alcohol use and ethnicity, and ethnocultural groups whose members drink yet rarely have problems may have things to teach other Canadians whose drinking gets them into trouble.

Religious and Other Beliefs

Many religious groups in Canada are opposed to alcohol use, including Hindus, Muslims, and fundamentalist Christians. The older, more established Protestant groups, such as Baptists and Presbyterians, have been joined by Mormons, Jehovah's Witnesses, and Seventh Day Adventists, as well as fundamentalists associated with other denominations. Some religions prohibit alcohol use among their followers, while others feature alcohol in religious ceremonies and take no stand on social drinking. Active members of religions that prohibit alcohol tend to abstain or drink infrequently and, in general, Canadians who frequently attend religious service drink less than others (see Table 5.1).

One study of the relationship between drinking and religious affiliation, feelings and behavior among students has shown these relationships to be complex (Adlaf and Smart, 1985). Those who were not affiliated with any religion were *less* likely to reporting drinking than those affiliated with Protestant or Catholic churches, but those who went to church most often and those who never went to church, had similar levels of drinking. Those who felt "very religious" reported drinking less than others. These seemingly contradictory findings are not easily explained unless many of the students classified as "non-affiliated" were actually adherents of non-Christian religions opposed to drinking.

Not much is known about the relationship between alcohol use and non-religious beliefs and practices (e.g., political affiliation). However, there are no good reasons to believe that abstainers and different types of drinkers can be distinguished by their views on most topics. Some relationship between beliefs about drinking and actual drinking behavior might nonetheless be expected.

A survey conducted in the Durham region of Ontario addressed this issue in some detail (Smart and Liban, 1981). Those interviewed were asked for their opinions on: (1) how much is acceptable to drink in different situations; (2) how, when, and how much males and females of different ages should drink; (3) the treatment of alcoholism; and (4) what should be done about individuals with various alcohol-related problems (e.g., wife abuse, public intoxication, driving while under the influence). Respondents were also asked about their own drinking and any related problems.

Most of the almost 1,000 people interviewed expressed "proper" opinions. Most did not approve of drunkenness, drinking and driving, or drinking while looking after children. Although many thought that it was sometimes all right to "drink enough to feel the effects but not to get drunk," this behavior was only considered acceptable at parties. Only nine per cent of those interviewed felt that it was all right for certain people to get drunk occasionally. Over 80 per cent felt that 16-year-olds should abstain completely. Most also felt that "something should be done" (by the police, relatives, or social service agencies) about such problems as alcohol-related wife abuse, public drunkenness, overspending on alcohol by heads of families, and drinking by bus drivers.

When attitudes and self-reported drinking behaviors were compared, they seemed reasonably consistent. Thus, for example, those who approved of heavy drinking also reported drinking more in the previous year and almost all those who thought it was all right to drink during working hours also reported feeling the effects of alcohol while on the job in the previous year.

Older males were the only group with an indifferent attitude toward the treatment of alcoholism. They felt that people with drinking problems should look after themselves. They also felt that the community had no responsibility to intervene and that treatments were ineffective.

Questions about attitudes to intoxication and to drinking were included in the *1989 National Alcohol and Drug Use Study* (Eliany, Giesbrecht et al., 1992). The results were generally consistent with those in the earlier Durham survey, with most people reporting quite conservative views. The majority (72 per cent) did not think that it was acceptable to "drink enough to feel intoxicated" in any social situations including in bars or at parties. Approval of intoxication was greatest among younger males and those who reported the highest levels of consumption. Only half of the males who drank more than 15 drinks a week disapproved of "drinking enough to feel the effects," even with other males in bars. Some implications of these results will be discussed in Chapter 11.

Who Drinks Most?

Thus far we have considered how specific social and economic factors influence drinking patterns. What happens if factors are combined, and which factors best predict heavy drinking? Eric Single and Joan Brewster, of the Canadian Centre on Substance Abuse, used data from the 1993 Canadian survey on alcohol and drugs to find the best predictors of heavy drinking from a set comprising age, gender, marital status, income, education, employment, religious attendance, language spoken at home, community size and region. The results showed that Canadians who are most likely to drink and to drink heavily tend to be young, English or French speaking, unattached males who are relatively well off and who rarely or never attend religious services (Single, Brewster et al., 1994).

Additional insight into the characteristics of Canadians who drink heavily comes from studies of young people. Many such studies have been conducted at the Addiction Research Foundation and have been summarized elsewhere (Smart, 1993). In brief, these studies show that some young people are especially prone to drink heavily and that they do so when alcohol is readily available. Availability is greatest for young people whose parents are heavy drinkers and who thus have plenty of alcohol around the home. In this situation, young people see heavy-drinking role models and may come to view heavy drinking as normal. Youth who drink heavily also tend to have friends who drink and who probably give them alcohol. Young heavy drinkers are more likely to have part-time jobs and thus to have more money to spend on drinking. They are also prone to use alcohol to deal with personal, family and school-related problems.

Youth who drink heavily are more anxious and depressed than their peers and more often feel rejected by their parents. They are less likely to have the protective factors associated with church attendance or an ethnic heritage that discourages heavy drinking. Further they are more likely to experience problems and have lower academic expectations. We will return to issues concerning the characteristics of heavy drinkers in Chapter 7, when we look more closely at Canadians with have serious alcohol-related problems.

Summary

Drinking habits vary with age, gender, occupation, marital status, ethnicity and religious beliefs. In general, women drink less than men and have fewer problems. However, many women do drink and are especially at risk for some types of problems. Drinking by young people is generally associated with parties and other social events and underage drinkers say they do most of their drinking away from home. Heavy drinking among youth is associated with anxiety, depression and feelings of rejection. Also, many young heavy drinkers do not have the protective factors associated with church attendance or a temperance heritage. About 20 per cent of male and five per cent of female university students report drinking more than 28 drinks a week, and heavy drinking is associated with lower grades. Drinking problems among the elderly appear more common that previously recognised and are expected to become more prevalent as the current population of drinkers age.

Professionals and skilled workers are more likely to drink than people in other occupations. However, among drinkers, the highest levels of consumption are found among those in blue-collar jobs. Among the heaviest drinkers are disproportionate numbers of young, unattached males who are relatively well off and who rarely attend religious services. Some occupational groups have a disproportionate number of heavy drinkers. Members of some ethno-cultural and religious groups have distinctive patterns of alcohol use, although there are signs that these patterns are changing and more research is needed. Canadians do not generally approve of heavy drinking, drunkenness and drinking and driving, but those who are more tolerant of heavy drinking are themselves more prone to heavy drinking and related problems.

Chapter 6

Drinking Among Aboriginal People

Approximately 1.5 million Canadian adults identify themselves as purely Indian, Inuit, or of mixed origin (Statistics Canada, 1989). In total these represent about six per cent of the Canadian population. Of these, only the Inuit and Indians have special status in the eyes of the federal government, and Indians only have special status if they are "registered." Registered Indians are members of bands that signed agreements with the British Crown and, in return for land, were granted certain privileges (e.g., the right to live on reserves, health and education rights). About 60 per cent of registered Indians live on reserves across the country, and the majority of Inuit live in special settlements in the north. The reserves and settlements range in size and accessibility. Some are virtual outposts with 50 or fewer people, while others are urban complexes with populations in excess of 6,000.

We have chosen to pay special attention to drinking problems among aboriginal people because they are generally believed to be widespread and often thought to reflect biological differences between aboriginal people and others. However, there are many gaps in our knowledge of drinking among aboriginal people and no simple explanations for the apparently high levels of drinking problems in some native communities. As for most aboriginal people, drinking patterns and problems are influenced by many factors and there is little to justify the need for any special theories that link "native" drinking to "native" characteristics. However, it seems likely that drinking among aboriginal people is heavily influenced by current social and political factors rooted in the historical relationships between indigenous peoples and the rest of Canadian society.

Drinking and Drinking Problems among Aboriginal People

In Chapter 1, we noted that many early European accounts of life in North America include graphic descriptions of drunkenness and disorderly conduct among native people. It is important to note, however, that there are also early references to Indians who refused alcohol because they did not like it or because they recognized its potency. Howay (1942) quotes Ross Cox, who wrote about the Indians of the Columbia River in 1833. "[They] entertain a strong aversion to ardent spirits which they regard as poison. They allege that slaves only drink to excess; that drunkenness is degrading to free men." Howay also quotes Von Langsdorff on the Tlingit: "Though they liked brandy very much, they rejected it because they see the effect it produces, and are afraid that, if deprived of their senses they should fall into the power of the Russians."

Many early accounts also show that Indians could drink in a well-controlled manner. For example, Captain George Vancouver describes a ship's dinner party involving three Indians: "After dinner they did not make the least scruple of partaking of our repast with such wine and liquors as were offered to them; though they drank very sparingly seeming to be well aware of their powerful effect" (Howay, 1942).

The "Indian drinking problem" was nonetheless a feature of pioneer life and attempts to prohibit Indians' access to alcohol was made by the French government, the Catholic church and some Indian chiefs. Concerns about alcohol problems among the Inuit have also been raised since the time they first gained ready access to alcohol.

Current media reports and statements made by native leaders and advocates suggest that alcohol-related problems are widespread and severe in many aboriginal communities and among indigenous people living in urban centres (Assembly of First Nations, 1988; Native Women's Association of Canada, 1981; National Association of Friendship Centres, 1985). Over 60 per cent of aboriginal people consider alcohol to be a problem in their communities (Statistics Canada, 1993). Widespread alcohol- and drug-related problems in aboriginal communities have also been reported by aboriginal people in treatment programs and by other native people. Many examples are given in a recent book on drinking in native communities by the Indian writer Brian Maracle (Maracle, 1993). This book profiles many Indian people who report that relatives, friends and other members of their communities often had drinking problems. In the book's introduction, Maracle indicates that he, his parents and most of his brothers and sisters were once "boozers" and that seven of his relatives died from drinking or in alcohol-related acts of violence. Maracle also notes that alcohol was responsible for "broken marriages, apprehended children, car accidents, fights, broken bones, cuts, scars, arrests and jail terms" involving members of his family.

High levels of drinking and related problems among native people are also evidenced by hospital and treatment centre statistics, although some frequently quoted figures are quite dated. For example, in 1981, 17 per cent of all alcohol-related hospital admissions involved aboriginal people, while 60 per cent of all illness involving aboriginal people was alcohol-related. Rates of admissions to addiction treatment centres for native people were 13 times the national average (Department of Indian and Northern Affairs, 1981). Further, a study of native patients attending a general practice clinic in an isolated community in northern Ontario showed that 27 per cent were "definitely alcoholics" and that another 20 per cent were "probably affected by alcoholism" (Merskey, Brant and Malla, 1988). Rates of admission to addiction treatment centres for indigenous people are also 13 times the national average (Department of Indian and Northern Affairs, 1981).

High levels of drinking can also be inferred from statistics on suicide, homicide and other deaths from causes frequently associated with heavy drinking or the abuse of other substances (e.g., deaths from poisoning, overdose, accidents, exposure or neglect). Significantly higher rates of death from these causes have been reported for native people than for other Canadians. Particularly striking are the high suicide rates among aboriginal youth. One study, conducted in Saskatchewan, showed that native children aged 14 or younger were 34 times more likely to commit suicide than other Saskatchewan residents in the same age group. For those aged 15 to 24, the suicide rate for indigenous people was over 10 times that for others (Fiddler, 1985).

Although there is only limited reliable information on the prevalence of alcohol-related birth defects (see Chapter 7) among babies born to native women, there is good reason to believe that, in some communities, the rates are much higher than for non-natives (Ashley, 1992). This is consistent with the notion that, in some native communities, the prevalence of heavy drinking among pregnant women is higher than in Canadian society generally.

Alcohol sales in communities with Indian reserves also indicate high levels of drinking among native people. This was shown in a study conducted in Ontario. Over the whole province, per capita alcohol consumption was 12.06 litres in counties with reserves but only 9.95 in other counties. In the North, per capita consumption levels were 13.42 litres for counties with reserves and 11.11 for others. These differences held up when income levels, employment and industrialization rates and other relevant factors were taken into account (Adrian and Williams, 1989).

Surveys of aboriginal communities are difficult to mount for a variety of reasons. Often these communities are very isolated, and many individuals refuse to co-operate in government-sponsored surveys because they are suspicious of the way the results will be used. Others refuse to participate in surveys because they feel that aboriginal problems have been studied for long enough and that the need is for more action and not for more research. Surveys may also be of limited value in communities that experience a high rate of heavy drinking, since drinking and its effects may themselves limit the ability or willingness of some individuals to participate. As a result, heavy drinkers are likely to be under-represented in the survey data, leading to an underestimation of overall drinking levels and the prevalence of heavy drinking.

A recent survey of drinking problems in native communities was conducted by Statistics Canada (Statistics Canada, 1993). This study also examined aboriginal languages, traditions, health and life styles and involved 135,600 Indians living on reserves. The individual response rate was good (77 per cent), but heavy drinkers were almost certainly under-represented among respondents. The value of the survey was also compromised by the fact that the residents of 78 reserves refused to be part of the study.

Among those surveyed, 67 per cent said that they sometimes drank alcohol. Of these drinkers, 60 per cent said that they drank less than twice a month. Only two per cent said that they drank every day. However, no information was collected on the amount drunk per occasion. Of those who said they did not drink, about half were life-time abstainers and the rest were former drinkers.

There were some quite large regional differences in the proportion of respondents who said that they currently drank alcohol. Those living in Newfoundland were the most likely to report current drinking (71 per cent) while those living in Prince Edward Island were the least likely to report drinking (54 per cent). The highest proportion of life-time abstainers was in Nova Scotia (22 per cent) and the lowest in British Columbia (11 per cent).

Some student surveys have compared alcohol and other drug use among native and non-native youth. For example, following a survey of high school students, 64 students who identified themselves as having native parentage were compared with 64 non-native students, matched by age, sex, grade level, geographic region, and father's occupation. The two groups of students reported very similar experiences with alcohol and drugs, and most reported no alcohol or drug-related problems (Liban and Smart, 1982).

A recent survey conducted in a small prairie city also found that native and non-native youth were equally likely to use alcohol. However, native youth were much more likely to use cigarettes, marijuana, non-medical tranquillisers, LSD and other hallucinogens. Also, in contrast with non-natives, there were fewer differences in the drug use patterns of native males and females (Gfellner and Hundleby, 1995).

One Quebec study compared students at a native school with French-Canadian students in rural schools (Lalinec-Michaud, Subak et al., 1991). The results showed few differences in drinking patterns except that the natives were *less* likely to report being regular users of alcohol. On the other hand, the native students were more likely to report that they had become violent after drinking and more likely to report excessive drinking among family members. Reported reasons for drinking were also different between the two groups. The main reasons for drinking given by the native students were pleasure (57 per cent), curiosity (27 per cent) and relief of depression (8 per cent). For the non-natives the main reasons were curiosity (39 per cent), relief of depression (29 per cent), pleasure (16 per cent) and to be accepted by others (13 per cent). Also, compared with others, the native students were more likely to report using amphetamines, hallucinogens and inhalants.

Clearly, there are many gaps in our knowledge of drinking among aboriginal people. While it appears that some native communities have high rates of alcohol problems, others do not. However, we have only limited information on the drinking patterns in different regions in different communities. More detailed information would help to identify communities with higher rates of abstinence or fewer alcohol-related problems. These could then be compared with communities where alcohol problems are more serious. Special attention should also be paid to factors such as urbanization, social organization and economic development (McKenzie, 1992).

Studies of the characteristics of low- and high-level drinkers in similar communities would also be of value in the identification of risk factors and the design of prevention programs. There are certainly many native people who drink without problems, but we know little about them or about the factors that distinguish them from those who drink heavily. Studies of non-native heavy drinkers suggest there are identifiable risk factors (see Chapter 7) but these may be different in native groups. Some variables that might be included in studies of native drinkers are: age, gender, tribal affiliation, place of residence and socio-economic status (Scott, 1992).

Biological Considerations and Drinking Problems among Aboriginal People

A common belief is that aboriginal people are genetically programmed to metabolize alcohol more slowly than others. This supposedly accounts for higher rates of heavy drinking and a tendency to become very intoxicated among aboriginal drinkers. However, direct support for the notion that native and non-native people react differently to alcohol is very weak. Some early, widely quoted studies used very small samples, failed to take account of subjects' drinking histories and made no attempts to determine if their subjects' ethnicity was correctly classified, except on the basis of their physical appearance (Fisher, 1987). More recent studies using improved methods have found no differences in alcohol metabolism rates between native and non-native people (May, 1989). Studies of liver enzymes involved in alcohol metabolism also fail to show differences between the two groups (O'Dowd, Rothhammes and Israel, 1990; Rex, Bosron et al., 1985).

Research on biological factors that may play a role in alcohol problems is continuing and it is now clear that individual differences in response to alcohol are partly genetically determined. It is possible that some "racial" groups, including some aboriginal people, will eventually be shown to have a genetic make up that makes them more vulnerable to alcohol. However, the differences between individuals will certainly be large. Moreover, we know that drinking patterns and rates of alcohol-related problems are heavily influenced by social and economic factors. In the case of aboriginal people, relationships with the dominant society and generally poor social and economic conditions certainly contribute to alcohol problems in some communities. Historical and present-day experiences with alcohol-control measures may also play a role in drinking problems among aboriginal people.

Social and Political Factors

Although Canadians who identify themselves as native people are quite heterogeneous, they share a history of colonization and cultural erosion. Having lost most of their lands, aboriginal people were sent to reserves or settlements. While such measures were taken ostensibly to maintain native cultures, it was often difficult or impossible to maintain traditional lifestyles in these settings. At the same time, missionaries preached against traditional beliefs, and churches and governments actively sought to undermine traditional customs and structures. Use of native languages was forbidden in many cases, and large numbers of native children were taken from their parents and sent to residential schools to be educated for menial jobs in white society. We now know that many of these children were physically and sexually abused by priests, teachers and foster parents. Whole generations of young natives were systematically taught that their cultures were inferior and that they should turn their backs on their heritage and join the rest of society.

However, native people have not always been welcomed in white society or equipped with the skills to cope with its expectations. Traditionally, many native children have dropped out of school and have lived in social and physical conditions that made it impossible to study or to develop self-respect. Housing and other conditions in many native communities are appalling and traditional institutions, including the native family, have been eroded or lost respect. Of course, many of today's young native people have also grown up surrounded by adults with alcohol or other drug problems.

Unemployment has always been very high in most aboriginal communities, but natives who leave reserves often lack the skills required to find satisfying jobs in the cities. White racism, loneliness, culture shock and other stresses of city life can compound existing problems and increase the risks for drinking and other problems.

Given these conditions it is not difficult to see the appeal of alcohol intoxication. For all its disadvantages, intoxication can provide an escape from the harsh realities of day-to-day life and bring a temporary feeling of worth. Intoxication may also be a way of expressing contempt for the broader society and of controlling anger at what that society has done to one's people (Lithman, 1979).

Northern resource development projects have also contributing to alcohol problems among native peoples (Giesbrecht, Brown et al., 1977; Hobart, 1978). The Berger (1977) inquiry into the proposal for the Mackenzie Valley pipeline and the Alaska Highway Pipeline inquiry found many telling examples of the impact of major development projects on northern native communities. These projects involved the influx of large numbers of young, unattached men earning relatively high wages while working in isolated areas under rough conditions. These men lived in work camps, trailer towns or boom towns. They often had little to do in their spare time and thus gravitated to local bars and women.

When local native labor was recruited for development projects, the workers earned unprecedented amounts, and much of their wages went on alcohol. Crimes of all kinds and general social disruptions increased. Berger stated quite starkly, "We can affirm one simple proposition: the more the industrial frontier displaces the home-land in the true North, the worse the incidence of crime and violence will be." Sadly he was correct.

Although many native people and their leaders are working hard to improve their situation, much remains to be done. There is good reason to believe that increasing degrees of self government and a revitalization of traditional customs and values will lead to material and other improvements and reduced drinking problems. However, because alcohol and other drug problems can undermine attempts to address other issues, they should also be tackled directly through prevention and treatment programs. Many such programs have been developed and it is likely that the most successful will include community development components and target whole communities, not just individuals who are at risk or those with problems.

Aboriginal Communities' Experiences with Alcohol

Most societies that allow people to drink have had hundreds, or even thousands, of years to experiment with various alcohol education and control policies. All drinking societies have also developed widely accepted formal and informal rules that affect the way people drink and behave after drinking. Even so, few societies seem completely content with present-day arrangements and many continue to experiment with ways to prevent or reduce drinking problems. This is even the case in France, which is often cited as having long-established and "civilized" drinking attitudes and policies. We have also seen that "mainstream" Canada has many alcohol-related problems. In Chapter 10, we will see that there is much that can be done to reduce these problems through education and thorough alcohol-control policies.

Aboriginal communities have had a relatively brief history of exposure to alcohol and, except for attempts at Prohibition, they have not generally experimented with alcohol-control policies. At the same time, many native people have only been exposed to the worst kinds of non-native drinking and have been victims of paternalistic policies developed by non-native governments.

Consider for a moment the way in which native Canadians were first introduced to alcohol. The first drinks they were given were very potent — rum, whisky or brandy. These drinks were provided in large quantities to people with no previous experiences with alcohol and no informal or formal rules to govern its use. Often alcohol was consumed during trade negotiations and white traders sometimes encouraged Indians to get drunk to undermine their ability to bargain. Further, many traders and others who had early contact with natives were themselves heavy drinkers whose habits were probably copied by the native people. Certainly, traders did not provide alcohol education or suggest alcohol-control measures with public health and safety objectives!

Some contemporary descriptions of drinking among early settlers are as lurid as those describing drinking among native people. Consider, for example, this account by the trader Daniel Harman: "Of all the people in the world, I think the Canadians, when drunk, are the most disagreeable; for excessive drinking generally causes them to quarrel and fight, among themselves. Indeed, I had rather have fifty drunken Indians in fort than five drunken Canadians." (MacAndrew and Edgerton, 1969).

And this, from the Jesuit Relations:

> But what makes their [soldiers stationed at Michilimackinac] misconduct on this score still worse is, that so persistent an attachment to the game [of gambling] is hardly ever unaccompanied by the general intoxication of all the players; and drunkenness is nearly always followed by quarrels that arise among them. When these occur publicly before the eyes of the savages, they give rise to three great scandals: the first at seeing them intoxicated; the second, at witnessing them fighting furiously with one another — sometimes to the extent of seizing their guns in order to kill each other, the third, at observing that the missionaries cannot remedy these evils. (MacAndrew and Edgerton, 1969).

The soldiers and traders who first had contact with native people were later followed by other "frontier" types with a traditional fondness for drink: loggers, trappers, oil-rig and pipeline workers (Hawthorn, Belshaw and Jamieson, 1957; MacKinnon and Newfeld, 1973; Gemini North, Ltd., 1974). We have also seen that heavy binge drinking was common among many non-natives in pioneer times and to this day, some native people in cities drink in run-down bars where many patrons drink heavily and sometimes get into fights.

At one time white governments banned the sale of alcohol to native people but this is no longer the case. However, many native communities are officially dry, following decisions taken by local bands. Many native leaders see no middle ground between alcohol-related mayhem and Prohibition. Local Prohibitions have, in some cases, lead to a significant reduction in drinking and related problems

(Smart, 1979a). However, elsewhere, Prohibitions led to increased bootlegging, home brewing, drug use and heavy binge drinking in town bars. This latter practice has resulted in a variety of problems including alcohol-related fights and accidents and deaths from exposure among intoxicated people falling asleep in the snow. Prohibitions in native communities have also resulted in residents gaining periodic access to large quantities of alcohol brought in by bootleggers. This can contribute to the "drink-it-all" parties typical of both skid-row and frontier drinking.

Native drinking problems may thus partly reflect the conditions under which drinking has taken place, both historically and currently, and the absence of culturally appropriate alcohol-control mechanisms with public health and safety goals. Examples of such control mechanisms could include agreed-upon times and places to drink, outlets and pricing policies that reflect local conditions, sanctions for violations, prohibitions against drinking by children, etc. However, because drinking problems in native communities both contribute to, and are exacerbated by, poor social and economic conditions, there are no quick or ready-made solutions. In the long run, native people must find solutions for themselves. Local prohibition may well be the best alternative for many communities. However, interventions imposed from outside and those without widespread community support are not going to work, and may make things worse.

Current Treatment and Prevention Programs for Aboriginal People

The federally-funded Addictions and Community Funded Program (ACFP) sponsors a variety of prevention and treatment programs for native people living on reserves. The ACFP was initiated in 1975 and was originally called the National Native Alcohol and Drug Abuse Program (NNADAP). In addition to programs for alcohol, drug and solvent abuse, the ACFP is responsible for a family-violence strategy involving prevention and treatment components. In 1993, the ACFP budget for addiction treatment was $3.8 million.

Provincial governments also provide some funds for native programs and native people can access all provincially funded addiction programs. Ontario has recently developed a strategy for native health concerns and ear-marked new funds for native addiction services.

The federal government funds 51 residential treatment programs for native people. They offer intensive non-medical treatment lasting four to six weeks. These programs tend to emphasize traditional native values, to use traditional treatment methods such as "sweat lodges" and to involve band elders and traditional religious leaders. Native programs are geared to abstinence and tend to incorporate elements of the Alcoholics Anonymous philosophy.

The majority of people admitted to these residential centres have problems with alcohol or with both alcohol and other drugs (especially narcotics and hallucinogens). About 40 per cent of admissions are women but the proportion of women increases from east to west. In the Atlantic provinces males outnumber females by a ratio of two to one, but in British Columbia about 40 per cent of admissions to native treatment centres are women. The majority of admissions are in the 25-34 year age group (Scott, 1992).

About two-thirds of those entering treatment complete the programs. This is similar to the rate reported for many non-native residential programs. Long-term outcomes have not been adequately studied and, as with treatment for non-natives, there is still much to be learned about what works and for whom (see Chapter 12). It is likely, however, that, as for non-natives, social and other environmental factors are major determinants of the outcome of treatment. Thus, the prospects for long-term recovery will be seriously compromised for those returning to communities with high unemployment rates and where many people drink heavily or use drugs. This is likely to be especially true for young people who return to families where adults and others drink and take drugs.

A recent example of successful treatment being compromised by social conditions attracted national attention. This involved a group of young inhalant abusers from the remote Innu community of Davis Inlet. These young people were taken *en mass* to a residential treatment centre in Alberta where they reportedly did well. However, within weeks of their return to a community plagued by alcohol, drug and other problems, many had relapsed.

The federal government also provides funds for addiction workers in aboriginal communities. They are involved in a variety of activities including counselling and referral. They also work with other health and social service providers on alcohol and drug education and prevention programs. A variety of prevention programs for native communities have also been developed by ACFP and its predecessor NNADAP. These include alcohol and drug educational programs, health promotion programs and initiatives that have a broader community development focus. However, the real impact of these programs has yet to be determined.

Summary

High levels of drinking and related problems are evident in some aboriginal communities, but there are clearly exceptions to this pattern and many gaps in our knowledge of native drinking patterns. Evidence to support claims that native people metabolize alcohol more slowly than others is very weak. However, historical factors, current social and economic conditions in many aboriginal communities, a lack of alcohol education and a lack of culturally appropriate alcohol-control policies seem to contribute to alcohol problems among native people.

Innovative treatment and prevention programs are being developed and some aboriginal communities have made great strides in dealing with alcohol problems through local prohibitions. However, to be successful, Prohibition strategies need the full support of community members.

Chapter 7

Alcohol Problems In Canada Today

Problems associated with the use of alcohol concern many Canadians and, from time to time, these problems take on importance as political issues or as the focus of media attention. Often the focus of concern is on problems of drinking and driving or on the need to strengthen relevant alcohol-control policies. The negative effects of drinking on health and behavior also receive a good deal of media attention, as do stories associating drinking with family violence and other crimes. In this chapter we will look at the alcohol-related problems that are of greatest concern to Canadians either because they may be personally affected or because these problems have a large impact on society as a whole.

Problems Associated with Alcohol Intoxication

To many Canadians alcohol intoxication is a problem because those affected behave in socially unacceptable ways. They may become noisy, aggressive, amorous or sentimental; they may fail to meet social obligations or pose a threat to themselves and others by driving cars or operating machinery. These are all legitimate concerns, but alcohol itself is only one contributing factor. The experiences and behaviors of people who drink small to moderate amounts of alcohol are quite variable. How drinkers feel and behave after a few drinks depends, among other things, on what they have been eating, their weight, their sex, their previous experiences with alcohol, how they feel before drinking and what they expect to happen. The behavior of apparently drunken people is also influenced by the behavior of others and by their expectations of how one should behave after drinking (drunken comportment). Thus relatively sober people sometimes appear quite drunk and some highly intoxicated people may appear quite sober in some situations.

Nonetheless, most drinkers show objective signs of impairment after two or three drinks or when their blood alcohol levels reach 20 to 30 mg of alcohol per 100 mL of blood. Reaction times are typically reduced at this level of intoxication and are further suppressed as intoxication increases. With increased levels of intoxication, the ability to judge time, distance, and speed is impaired, as is physical co-ordination. Most drinkers will appear quite drunk at blood alcohol levels in excess of 150 mg/100 mL of blood. Although many Canadians sometimes get intoxicated, the behavior of intoxicated people is often considered irritating or offensive. Thus, 21 per cent of Canadians report that in the past year, they have been insulted or humiliated by someone who had been drinking (Single, Brewster et al., 1994).

Intoxication and criminal liability became the subject of media attention and heated public debate following a ruling by the Supreme Court of Canada in September 1994. In its ruling, known as the Daviault decision, the court allowed an appeal of a sexual assault conviction and ordered a new trial on grounds that the accused was entitled to use the defence of extreme alcohol intoxication — a defence that had been denied by a lower court. At issue in the case was whether Daviault had been able to form the mental intent necessary to convict him of a "general intent" crime like sexual assault, or whether extreme drunkenness had put him in a state akin to "automatism," and therefore absolved him of guilt. In its ruling, the court upheld Daviault's right to use evidence of his intoxication — which suggested his blood alcohol level was between 400 and 600 mg/100 mL, a level that can lead to coma or death in inexperienced drinkers — as a defence.

Although the court had indicated that it did not wish to open the door for large numbers of defendants to argue that they were "too drunk to know better," this soon occurred in at least three other criminal cases. However, protest by women's groups and expert challenges to the concept of alcohol-induced automatism (Addiction Research Foundation, 1995) led the solicitor general to introduce legislation disallowing the use of self-induced intoxication as a defence in general-intent crimes. Known as Bill C-72, the legislation was passed in June 1995.

Drunkenness per se is not a criminal offence, although in all parts of Canada, it is an offence to appear drunk in a public place. At one time public drunkenness was a major problem in many Canadian cities and police spent a good deal of time arresting drunks and holding them in "drunk tanks" in police cells. However, the number of arrests recorded for intoxication has generally decreased in the past 20 years. Reasons for this decline include the development of detoxification and rehabilitation programs, improvements in the welfare system, deaths among older "skid row" alcoholics, and a general disillusionment with dealing with public drunkenness through the revolving doors of the criminal justice system (Ogborne, Adrian et al., 1991).

Hangovers and Other Minor Problems

Many of the congeners, or additives, used to give taste and aroma to alcoholic drinks have toxic effects. When combined with the dehydration caused by the alcohol itself, the results can include depressed mood, headaches, nausea, and sensitivity to noise — a condition that most drinkers refer to as a hangover. Hangovers vary in severity and duration and are a function of the type and amount of drink consumed. Vodka, with few additives, tends to result in mild hangovers, whereas red wine hangovers are much more severe because of its high concentrations of congeners. Hangovers are not particularly serious in themselves, but they certainly account for a good deal of lost productivity as a result of lateness, absenteeism and reduced work performance. Among university students, almost 50 per cent report experiencing hangovers, while 13 per cent report missing classes due to hangovers (Gliksman, Adlaf et al., 1994). Those suffering from hangovers are also less competent as drivers than those who have completely sobered up (Frank, 1983).

Occasional, mildly intoxicating amounts of alcohol do not appear to cause significant medical problems in healthy adults. Gastric upsets do occur in some people, which may predispose them to stomach or duodenal ulcers. Drinking larger amounts increases the risk of gastritis and can also lead to diabetic coma or to another serious condition known as acute pancreatitis (Korsten and Wilson, 1993).

Blackouts

"Blackouts" are episodes of memory loss experienced by chronic heavy drinkers and by some social drinkers who drink far more than usual. Those affected have no recollection of events that occurred sometime before and during an episode of drinking. For example, they may wake up in a strange place and have no recollection of how they got there. Blackouts have been suspected of heralding the onset of chronic alcoholism, but studies show that 20 to 30 per cent of light to moderate drinkers have had at least one episode of blackout (Kinney and Leaton, 1995).

Blackouts seem to occur because alcohol interferes with the brain's ability to convert short-term to long-term memories. This would account for the fact that people who experience blackouts appear to function quite normally at times that they are later unable to recall. At these times long-term memories are unimpaired and remain intact after recovery from the blackout (Ryan and Butters, 1983).

Traffic Accidents and Offences

Alcohol intoxication is a factor in many motor vehicle and pedestrian accidents. This is evident from recent data published by researchers at Traffic Injury Research Foundation in Ottawa and summarized in the *Canadian Profile* by Single, Williams et al. (1994). These data show that, in 1990, 35 per cent of fatally injured car drivers had a blood-alcohol content (BAC) in excess of the legal limit of 80 mg/100 mL and that 53 per cent had been drinking before their deaths. Fatally injured drivers in the 26-35 age group were the most likely to have been drinking and to have BACs in excess of the legal limit. Most of those killed in alcohol-related traffic accidents (92 per cent) were males and 62 per cent of accidents involving drinking drivers were single-vehicle accidents. In contrast, only 24 per cent of crashes where the drivers had not been drinking involved single vehicles. Other studies show that high proportions of fatally injured tractor/trailer drivers (23 per cent), truck/van drivers (52 per cent), motorcyclists (54 per cent), and pedestrians (23 per cent) were alcohol impaired, or had been drinking, prior to death.

The number of people involved in alcohol-related traffic accidents has, however, been dropping steadily since 1974, when national statistics were first kept. From 1984 to 1991 the number of fatal accidents involving drinking drivers fell from 492 to 318 (a 35 per cent reduction), and the number of non-fatal accidents fell from 23,247 to 14,493 (a drop of 38 per cent). The blood-alcohol levels of those involved in fatal crashes has also declined over the last decade. It is hoped that this is due to changing attitudes and may reflect the positive effects of police spot checks and other features of Reduce Impaired Driving Everywhere (RIDE) programs.

However, concerns about drinking and driving among youth remain high. Traffic accidents are the leading cause of death among people aged 15 to 24, resulting in 42 per cent of deaths among males and 36 per cent of deaths among females in this age group. Many of these accidents are alcohol related. Recently there have been several deaths involving young people returning from "bush parties." These parties have a long tradition in Ontario. They are usually held in the spring and summer and many involve informal gatherings of large numbers of young people on rural properties. Of course, most guests arrive in motorized vehicles and some drive home after drinking at the parties. The number of alcohol-related accidents following bush parties is unknown. However, the risks are clear and, according to police, both organized and informal bush parties result in many drinking drivers.

Although most Canadians are concerned about drinking and driving, and support stiffer penalties, many also admit to drinking and driving themselves. Thus in one recent survey, 20 per cent of respondents reported that they had driven after drinking in the past month (Smart, 1993). Self-reported driving after drinking was particularly common among males, those between 25 and 45, those with more education and higher self-reported levels of alcohol consumption. Among students, the proportion who reported drinking and driving in the previous year in a 1995 survey was 24 per cent, up slightly from 1990 (20 per cent) but still only half as many as those reporting drinking and driving (58 per cent) in 1977 (Adlaf, Ivis et al., 1995).

Other Accidents

Alcohol is implicated in many drownings, accidental falls and accidents involving fire (Walsh and McLeod, 1983; Cherpitel, 1994). Alcohol and drug use are also often blamed for accidents in the workplace. However, a recent Ontario survey indicated that work-related accidents were more likely to be associated with poor working conditions and shift work than with alcohol or other drug use (Macdonald, 1995).

Overdose and Poisoning

Very large doses of alcohol can lead to alcohol poisoning and unconsciousness. This can occur with adults who drink until they pass out, but some cases of alcoholic poisoning involve children who unwittingly drink alcohol or who are "experimenting." In 1989, 985 people were admitted to hospital suffering from the toxic effects of beverage alcohol. Of these, 36 per cent were under 20 years of age and 22 per cent were under 14 (Addiction Research Foundation, 1994).

Assault, Rape, Murder and Other Crimes

The recent debate about drunkenness and criminal responsibility sparked by the Supreme Court of Canada's 1994 ruling in the Daviault case has been very much focused on crimes of violence. Many Canadians are especially concerned that men may seek to use drunkenness as an excuse for violence against women. This would be a tragedy, because there is no doubt that alcohol and violence are linked in our society. In the general population 5.6 per cent report being assaulted by someone who had been drinking (Single, Brewster et al., 1994). Among families whose domestic problems come to court, violence is more than twice as common in families where a member has an alcohol problem than in families without alcohol problems (Langevin, 1982). Alcohol is consumed by the aggressor (usually male) in 50 per cent of cases of spouse abuse, in 30 per cent of cases of child abuse, and 50 per cent of homicide cases. At least 60 per cent of rapists are also believed to have been under the influence of alcohol at the time of the assault (Gerson and Preston, 1979; Johnson, Gibson and Linden, 1978) as are 30 to 50 per cent of all sex offenders (Barbary, undated).

The relationship between alcohol and violence is, however, complex. This is evident from a recent study conducted in Thunder Bay, Ontario (Pernanen, 1991). The study found that, over a one-year period, police records showed that over 42 per cent of all violent crimes involved alcohol consumption by the offender, the victim or both. Further, a victimization study showed that alcohol was involved in over 50 per cent of violent day-to-day confrontations (threats, pushing and shoving). However, drinkers observed in bars were twice as likely to kiss and hug than to behave aggressively. This suggests the behaviors while drinking are influenced by settings and expectations as well as by the direct effects of alcohol. It also shows that saying "the drink made me do it" is a poor excuse for any type of behavior while intoxicated.

Many non-violent crimes are also committed by people under the influence of alcohol. Often, those involved use alcohol to give them courage or find themselves more adventuresome when they have been drinking. A high proportion of unsuccessful criminals (those who are caught and sent to jail) have drinking problems and thus tend to have been drinking when engaged in criminal activities.

Adverse Reactions between Alcohol and Other Drugs

Alcohol enhances the effects of many other drugs including marijuana, antihistamines, tranquillizers, barbiturates and other sleeping pills (Kissin, 1974). Many accidental deaths have been attributed to mixing alcohol with these drugs, but data on the number of such deaths and on other problems concerning the mixing of drugs and alcohol are not readily available.

Adverse reactions involving alcohol and drugs among elderly Canadians are of particular concern. Some elderly people are especially vulnerable to such reactions because of lower body weights, poor health and the use of multiple prescription drugs (Williams, 1984). Frequent use of various drugs is also common among heavy drinkers and these "mixed" drug and alcohol abusers generally have more problems than those who only use alcohol (Ashley, le Riche et al., 1978).

Fetal Damage and Problems in Pregnancy

It has long been known that heavy drinking during pregnancy poses risks to unborn children. For example, in ancient Carthage there were rituals against drinking on wedding nights to prevent birth defects (Jung, 1994). Reports of physical and mental problems among children born to alcoholic mothers have appeared in scientific literature for many years. However, it is only in the past 20 years or so that alcohol-related birth defects have been the subject of much research.

Fetal problems shown to be associated with drinking during pregnancy range from subtle neurological deficits in some children to gross physical and mental problems in others. Heavy drinking during pregnancy can also increase the risk of spontaneous abortion. The term Fetal Alcohol Effects (FAE) has been coined to refer to a variety of mild to moderate consequences of drinking during pregnancy including low birth weight, hyperactivity, learning and motor difficulties and some physical defects. A cluster of more severe problems that include slow growth, small head, facial deformities, heart irregularities, and mental retardation has been labelled the Fetal Alcohol Syndrome (FAS), which has only been identified in children born to mothers who drink heavily during pregnancy.

The exact mechanisms by which alcohol damages the fetus are uncertain and it is not always possible to rule out the effects of other factors such as smoking, poor nutrition and poor health. However, recent studies involving rats suggest that a variety of mechanisms are involved. These include a reduction in the transfer of nutrients across the placenta and toxic effects of acetaldehyde. Most likely, different mechanisms are involved at different stages of pregnancy and contribute to different fetal abnormalities (U.S. Department of Health and Human Services, 1993).

Not all women who drink heavily during pregnancy have children with FAS or FAE and research is being undertaken to identify factors that increase the risk of alcohol-related birth defects. Genetic, nutritional and environmental factors seem to play important roles and researchers are trying to develop screening tests to identify high-risk mothers.

One Canadian study of mothers who drank during pregnancy failed to show a relationship between mothers' drinking and neurobiological functioning in their children (Fried and Watkinson, 1988). However, in this study most of the "heavy drinkers" consumed less than one drink per day. Other studies involving heavier drinkers have generally shown positive relationships between maternal drinking levels and physical birth defects, growth deficiencies, mental and behavioral problems (DHHS, 1993). Although questions concerning "safe limits" for drinking during pregnancy have not been finally resolved, the United States surgeon general and Canadian addiction experts have advised pregnant women to abstain completely.

There are few Canadian statistics concerning alcohol-related birth defects. However, of children with physical or learning problems in Northern British Columbia and Yukon, it has been suggested that 30 per cent could be characterized as FAE or FAS children. Most of these children were from First Nations communities where alcohol-related problems were very common. Alcohol-related birth defects are generally recognized as a major problem among native people in North American and native organizations are currently very concerned about alerting women to the dangers of drinking during pregnancy (National Native Advisory Committee on Alcohol, 1983).

Studies done in the United States have suggested that the prevalence of Fetal Alcohol Syndrome ranges from 0 to 1.58 per 1,000 live births, depending on ethnic background and socio-economic status. Prevalence rates for FAE are estimated to be three times higher and some authorities believe that alcohol damage to the fetus is a major cause of mental retardation in the Western world (Graham-Clay, 1983).

Problems Associated with Chronic Use of Alcohol

We know about the effects of long-term alcohol consumption from animal studies, from follow-up studies of people drinking at different levels, studies of drinking histories among people suffering from various diseases, and studies comparing users of varying amounts of alcohol with abstainers. These studies show the kinds of problems that can occur after long-term, heavy-drinking alcohol use and also suggest what levels drinking may be generally safe.

Alcohol Dependence

People who regularly use alcohol or other mood-altering drugs may begin to experience feelings of dependence or behave in ways that earn the label "drug dependent." For example, they may go out of their way to ensure a regular supply. Regular users may also have physical symptoms when drugs are withdrawn. Such dependent users are commonly referred to as "drug addicts" or "alcoholics" if their preferred drug is alcohol.

Experts in the field of psychoactive drug use do not agree on the definition of the terms "dependence" or "addiction" and especially "alcohol addiction" and "alcoholism." Many consider all of these term too vague and charge those who use them with misleading attempts to simplify complex, variable phenomena. Others seem to use the terms unapologetically and have no doubts that they refer to distinct conditions characterized by objective symptoms. Finally there are those (probably the majority) who tend to avoid use of the terms "addiction" and "alcoholism" and talk about drug users as exhibiting varying degrees of "dependence" (e.g., Skinner and Allen, 1982). Those who favor this approach point out that among users of most drugs there are those who can be considered "mildly dependent," that is, users who experience feelings of loss when the drug is not available and show concerns that they always have a steady supply. However, the distinction between these "mild" cases and those who are "severely dependent" is seen as a matter of degree. In other words, alcohol and other drug problems are seen as existing along a continuum and not as a distinct condition that can be precisely diagnosed.

Increasing dependence on alcohol is evidenced by a preoccupation with drinking, a strong desire to drink, a sense of "loss of control," drinking to avoid withdrawal symptoms, high levels of drinking despite a variety of negative consequences, increased tolerance to the effects of alcohol and relapse after abstinence. Withdrawal symptoms among severely dependent people include tremors, anxiety, gastric distress, headache, fever, sweating and sometimes convulsions and hallucinations.

Both the World Health Organization (WHO) and the American Psychiatric Association (APA) have developed criteria for "alcohol dependence" (World Health Organization, 1992; American Psychiatric Association, 1994). These two sets of criteria are similar, as are the criteria used to define "harmful use" (WHO) and "alcohol abuse" (APA). In a recent study of adults in Ontario (Adlaf, Ivis and Smart, 1994), 5.3 per cent of respondents met the WHO criteria for alcohol dependence, reporting three or more of the following six experiences in the previous 12 months:

1. A strong desire to use
2. Difficulties in controlling use
3. Physiological withdrawal or use to relieve withdrawal
4. Tolerance (need to increase dose to obtain effects)
5. Progressive neglect of other activities in favor of use
6. Persistent use despite adverse physical or mental effects

Estimates of the number of Canadians heavily dependent on alcohol ("alcoholics" in the popular sense) have also been made using various formulas involving alcohol sales data, liver-cirrhosis death rates, deaths attributed to alcoholism, and alcohol-related suicides. These procedures give widely different results. The most popular formula, based on death rates for liver cirrhosis, resulted in the estimate that there were 486,100 alcoholics in Canada in 1989 — about three per cent of the population aged 15 and over (Single, Brewster et al., 1994).

Drinkers seeking help for alcohol problems are often severely dependent on alcohol and have many related problems. They are thus likely to be considered "alcoholics" by those who still use this term. However, within the treatment population there are people with widely varying levels of alcohol dependence (Skinner, 1979). Some show few signs of dependence and may not even drink excessively. Rather, they are concerned about problems such as driving-while-impaired charges. Others have mild to moderate levels of dependence on alcohol, but do not experience physical withdrawal symptoms. They are nonetheless beginning to experience social and psychological problems as a result of their drinking and see the need to cut down. Some therapists believe that such people can learn to moderate their drinking, and there is good evidence that many can (Miller, 1983). However, most experts agree that heavily dependent drinkers should be encouraged to abstain completely and permanently.

Liver Disorders

Long-term heavy drinking can have serious consequences for almost every organ and system in the body. Some people are quite vulnerable to these problems especially when they have other health problems or poor diets.

Among Canadians seeking assistance for alcohol-related problems, liver disorders of various kinds are fairly common. For example, 46 per cent of patients admitted to a specialized inpatient unit in Toronto had fatty livers (Ashley, Olin et al., 1981). Liver disorders, except for cancer, are not common among those who are not alcohol-dependent. Up to 73 per cent of all Canadian deaths from liver cirrhosis (2,155 in 1990) are probably related to alcohol consumption. Changes in liver cirrhosis death rates in the general population correspond closely to changes in per capita alcohol consumption and alcoholism (Schmidt, 1977).

Heavy drinkers with liver damage can prevent further deterioration by abstaining and, in young people, the liver has the capacity to grow new tissue to replace dead cirrhotic cells. However, this capacity diminishes with age and consequently new treatments for alcoholics, liver damage are being studied. Researchers at the Addiction Research

Foundation have shown that alcoholics with liver disease have reduced death rates if they are given the drug propylthoriouracil (PTU) (Orrego, Blake et al., 1987). However, this does not work if the liver is too badly damaged. Liver transplants may be required in such cases and have proven to be quite successful (DHHS, 1993).

Neurological Disorders and Alcohol

It is well established that when used excessively over long periods, alcohol itself damages the human brain. Among some heavy drinkers, alcohol-related brain damage is compounded by damage caused by vitamin deficiency. Alcohol-related brain damage can be detected with psychological and neurological tests and with special computer-analysed X-rays that can detect evidence of tissue loss and other anatomical changes (Carlen, Wilkinson et al., 1981). People suffering alcohol-related brain damage have difficulties concentrating, solving problems and learning. In severe cases they will appear demented — a condition known as alcoholic dementia. About five per cent of chronic alcoholics have some form of chronic brain damage (Ashley, Olin et al., 1981). The two most common forms, Wernicke's encephalopathy and Korsakoff's psychosis, lead to severe disturbances in vision, balance, and thinking and require skilled management and hospitalization.

Some mental capacities are more impaired by alcohol than others, leading some researchers to speculate that alcohol may damage brain structures selectively. Studies using tests sensitive to frontal lobe or right hemisphere damage tend to support this view. An earlier notion (Jones and Parson, 1977) that alcohol-related brain damage was similar to the organic deterioration that occurs with normal aging has largely been discounted. This is partly because alcoholics show improvements on tests of mental abilities with increasing abstinence (Ryan and Butters, 1983).

Some researchers have suggested that long-term social drinkers may also suffer some degree of alcohol-induced mental impairments. However, the evidence is by no means clear cut and relevant studies have not always controlled for intelligence, education and other factors that influence scores on the test used (Delin and Lee, 1992). The issue is nonetheless of great importance because many people drink socially over very long periods. Research is thus continuing in this area.

Mention should also be made of a condition known as alcoholic psychosis. Those affected lose the ability to distinguish reality from fantasy. They may have irrational fears and wild fantasies, particularly regarding plots to kill or harm them. In 1990, 5,437 Canadians were admitted to hospital with a primary diagnosis of alcoholic psychosis (Addiction Research Foundation, 1994).

Heart Disease, Cancer and Other Health Problems

There is some evidence that consuming small amounts of alcohol reduces the risk of heart disease in some drinkers (see below). However, heavy drinking certainly increases hypertension, weakens heart muscles and leads to disturbances in heart rhythms. The risk of stroke also increase with heavy drinking (MacMahon, 1987). Alcohol consumption has shown to be a causal factor in cancers of the mouth, pharynx, larynx, esophagus and liver (International Agency for Research on Cancer, 1988). There is also evidence that alcohol use increased the risk of lung cancer (Potter and McMichael, 1984) and breast cancer among premenopausal women (Friedenreich, Howe et al., 1993).

A recent Ontario study found positive correlations between per capita alcohol consumption and rates of death from cancer of the lung, mouth, esophagus, intestines, larynx and breast (women only) for the period 1963-1983 (Anglin, Mann and Smart, 1995). However, correlations between cancer death rates and per-capita consumption were generally small and were insignificant for the period between 1975 and 1983, when per-capita alcohol consumption declined. This may be due to a time lag between the onset of heavy drinking and death from alcohol-related cancer. Chronic alcohol use also depresses the immune system and leaves the drinker vulnerable to a variety of infections. Those with HIV or AIDS should thus be very moderate in their use of alcohol or abstain completely.

Canadians seeking help for alcohol-related problems tend to suffer form sexual impotence, sleep disturbances, anaemia, shrinkage of the testicles, gout, depression, diabetes, chronic pancreatitis, epilepsy, and various skin disorders. Heavy smoking, which is often part of heavy drinkers' lifestyles, contributes to these disorders and to infectious diseases such as pneumonia, tuberculosis, and bronchitis (Ashley, Olin et al., 1981).

People suffering from conditions that are not themselves alcohol-related may also find some problems exacerbated if they drink (e.g., heart problems, diabetes). Chronic intoxication can also exacerbate symptoms of Parkinson's disease and a variety of other problems. Further information on these problems can be found in articles published in a special edition of the journal *Alcohol Health and Research World* (Vol. 17(4), 1993), concerning lesser-known medical consequences of alcohol use.

Premature Deaths and Drinking

Heavy drinkers are clearly at risk for a number of medical disorders that can shorten their lives. These risks include those directly associated with the effects of alcohol as well as those associated with related lifestyles (e.g., smoking and poor nutrition).

Five Canadian studies have compared the death rates of alcoholics with those of people of a similar age (deLint and Levinson, 1975). These studies suggest that alcoholics are two to four times as likely to die within any given period than are non-alcoholics of the same age. In 1990, 2,959 deaths in Canada were officially recorded as directly related to the consumption of alcohol. The majority (72 per cent) involved alcohol-induced liver disorders. An estimated additional 15,665 deaths were alcohol-related, including 1,805 deaths in motor vehicle accidents. However, because alcohol can cause or exacerbate so many disease processes, its total contribution to premature death among Canadians is probably far greater than these figures suggest. Some experts believe that as many as one death in every 10 is related to alcohol in some way (Health and Welfare Canada, 1984).

Suicide and Other Problems

Both acute intoxication and chronic alcohol use are important factors in many suicides. Studies in Canada and the United States have shown that the proportion of suicides with alcohol in their blood ranges from 20 to 50 per cent, depending on the sample and the completeness of the autopsy information (Coade, 1974; James, 1966; Frankel, Ferrence et al., 1976; Saskatchewan Alcohol and Drug Abuse Commission, 1988). The Saskatchewan study also found that 18.8 per cent of suicides had blood levels of 80 mg/100 mL or higher at the time of death.

Alcoholism has been characterized by some researchers as a form of suicide (Menninger, 1938) and the estimates of actual suicide among alcoholics have typically been higher than in the non-alcoholic population (Edwards, Kyle et al., 1978; Schmidt and deLint, 1972; Sundby, 1967). The proportion of alcoholics among suicides has been estimated to range from six to 30 per cent (Beck, Weissman and Kovacs, 1976; Mayfield and Montgomery, 1972; Murphy, Armstrong et al., 1979), and suicide rates were highly correlated with per capita alcohol consumption during the period of increasing consumption in Ontario between 1975 and 1983 (Smart and Mann, 1990).

Heavy drinking over a long period is often associated with a variety of other mental health and social problems. In some cases these predate the onset of alcohol problems, but often they are caused or exacerbated by heavy drinking. Common problems include loss of friends and other relationships, loss of employment or poor work performance, economic problems and trouble with the law. Many chronic heavy drinkers are also depressed.

It is not only the individual drinker who is affected by these problems. Over 40 per cent of Canadians report that they have been bothered by someone else's drinking in the past year (Single, Brewster et al., 1994) and it has been estimated that every chronic alcoholic has a significant negative effect on up to 15 other people, whether family members, co-workers or friends (Ehline and Tinghe, 1977).

Who Has the Most Problems with Alcohol?

We have seen in this chapter that alcohol contributes to many types of problems, some involving large segments of the population. In fact, over nine per cent of Canadian drinkers report that drinking has had an adverse effect on their social life, physical health, happiness, home life, work or finances in the past year (Single, Brewster et al., 1994). In some groups, such as male university students, the majority report experiencing some types of alcohol-related problems. However, it is obvious that some people experience more problems than others and that, for a minority, their lives are dominated and completely ruined by their drinking.

Who are the people with the most problems? Do they have anything in common except for alcohol-problems? Why do they drink despite negative consequences? The latest Canada-wide survey gives use some answers to these questions. As we have already noted, this survey showed that the heaviest drinkers were young, unattached, relatively well off, English- or French-speaking males who infrequently attended religious services. A similar picture emerges when the survey data is used to explore the characteristics of people with alcohol problems. They tend to be male, single or divorced and to attend religious services infrequently. However, unemployment and/or relatively low income are also significantly related with drinking problems.

Of course, the similarity between the portrait of heavy drinkers and that of problem drinkers is hardly surprising. Frequent heavy drinking increases the chances of problems associated with intoxication and problems linked with chronic use. However, this is not the whole story. Some people drink heavily without problems, while others experience problems at relatively low levels of consumption. Also, more detailed studies of people with more serious problems suggest that their characteristics are quite varied. Nevertheless, two questions about possible common characteristics among chronic heavy drinkers continue to intrigue researchers and the general public. The first concerns the existence of an "alcoholic personality", while the second concerns the influence of genetic factors on drinking behaviors.

There is a long history of research concerning the "alcoholic personality." This has been based on the assumption that alcoholics have common traits that cause them to drink as they do. Some of this research has looked for evidence that alcoholics have unresolved conflicts originating in childhood and are "orally fixated," "dependent" or "immature" (McCord and McCord, 1960). Other studies have looked for evidence that alcoholics have a frustrated need for power (McClelland, Davis et al., 1972). These theories reflect the influence of psycho-analysis. However, they do not lend themselves to rigorous testing because they concern early childhood events that are difficult to verify. Also, like other personality theories, they can be criticised for assuming that similarities between the personalities of adult alcoholics reflect causes rather than consequences of heavy drinking careers.

Many studies of people in treatment for alcohol problems suggest that they have some features in common. However, many of these common features are the result of years of heavy drinking and associated problems, and cannot be used to explain why heavy drinking occurs in the first place. Studies of people in treatment programs are also limited in what they tell us about heavy drinkers in the general population because very few people with drinking problems ever seek help from specialized programs (see Chapter 12). Those who do seek help are probably atypical in as much as they seem to have more severe problems and fewer social supports than others (Weisner, 1987).

One study of people in treatment suggests that, apart from having drinking in common, they collectively exhibit symptoms similar to those of people being treated for psychiatric problems: depression, neuroses, psychoses, anti-social tendencies, phobias, or other problems that typically lead people to seek psychiatric help (Skinner, Reed and Jackson, 1976; Ross, Glasser and Stiasny, 1988). American population studies suggest that people diagnosed as alcoholics are more likely to have psychiatric problems than others and that the diagnosis of anti-social personality disorder has the strongest association with alcohol problems (DHHS, 1993).

Research into the role of genetic factors in differences in drinking habits has produced intriguing but inconsistent results, as noted in Chapter 5. Most relevant studies have been done in the United States and in Scandinavia, and because this is a book specifically about Canada, we will not review these studies in detail. However, we will summarize the results of studies that lend most support to the view that genetic factors contribute to chronic alcohol dependence.

Several studies have shown that rates of alcoholism among identical twins are more alike than those of fraternal twins (Kaprio, Koskenvuo et al., 1987; Heath, Jardine and Martin, 1989; Pickens, Svikis et al., 1991). Because identical twins have the same genetic make up, while all pairs of twins have similar environments, these studies suggest that alcoholism at least partly genetically determined. Studies of twins separated at birth provide especially convincing evidence for the influence of genetic factors (Goodwin, Schulsinger et al., 1973, 1977; Bohman, 1978). Most of these studies were done in Denmark and Sweden where excellent birth and other registries provide rich sources of information for genetic research. These Scandinavian adoption studies show higher concordance for alcoholism among identical than non-identical twins. One recent U.S. study has since produced similar results (Kendler, Heath et al., 1992).

Studies of twins, especially adoption studies, have been criticised for the definitions of alcoholism used and for other methodological problems (Serles, 1988). It has also been pointed out that adopted children and their birth and adopted families may be atypical in several respects. For example, birth mothers are likely to have been unmarried teenagers at risk for having low-weight babies. However, most researchers agree that these studies provide fairly strong evidence for the influence of genetic factors, especially among males (DHHS, 1993).

Strong support for genetic influences on drinking levels and drinking patterns was provided by a recent study involving Asians living in North America (Tu and Israel, 1995). Researchers found that Asian men who did not carry a gene involved in metabolism of acetaldehyde (a toxic metabolite of alcohol) drank only one-third as much alcohol (six versus 18 drinks per week) than those with the gene. Those without the gene were also less likely to report drinking in binges (15 versus 42 per cent) and more likely to say that they were complete abstainers (39 versus 13 per cent) than those with the gene. While this study also found drinking levels to be lower among males who rated themselves culturally more "Oriental" than "Canadian/American," these differences were far less marked than differences associated with the presence or absence of the marker gene.

The study found no genetic or cultural influences on the drinking of Asian women. The authors suggest that this may result from different drinking norms among men and women. Men are generally expected to drink more than women and are thus more likely to experience negative reactions associated with genetically determined metabolic processes.

The ways in which genetic factors influence drinking behavior is, however, generally recognized as quite complex, and many different genes are probably involved. Areas of research include studies of genetically-determined differences in enzymes involved in the metabolism of alcohol, studies of enzymes that affect moods, and studies of differences in brain activities between the sons of alcoholics and others. Reviews of these and other relevant studies are given in a recent report prepared for the U.S. Congress (DHHS, 1993).

Of course, even if genetic factors influence drinking and related problems, this does not mean some drinkers are "doomed" to alcoholism because of their genetic make up. On the contrary, there is clear evidence that drinking and its consequences are influenced by environmental factors (Ogborne, Sobell and Sobell, 1985) and that individual drinkers learn to drink under the influence of others. While some people disregard clear signs that they may be vulnerable to alcohol problems because of a family history or a high tolerance, others take care not to drink too much lest they too develop problems. Drinking, like most human behaviors is too complex and variable to be explained by genetics alone. Our enthusiasm for genetic explanations should thus be tempered by attention to the effects of negative life experiences and other environmental factors.

Is There a Safe or Even Beneficial Level of Drinking?

Based on an extensive review of the international research literature, experts from the Canadian Centre on Substance Abuse and the Ontario Addiction Research Foundation have advised Canadians to limit their drinking to two or fewer drinks per day and to have a least one day of abstinence per week (Ashley, Ferrence et al., 1994). However, lower levels of consumption are advised for people of small stature, especially women. Pregnant women are advised to abstain completely, as are those with certain psychological and physical illnesses, those taking certain medications, people under the legal drinking age and those who have shown a persistent inability to control their drinking. Abstinence is also recommended for those operating machinery or vehicles and, while they are on duty, for those responsible for public order.

The same experts also examined the scientific literature for evidence that moderate drinking might lower the risk of coronary heart disease (CHD) or be associated with other health benefits. They noted that some American and European population studies have found a "J"-shaped relationship between drinking and rates of heart disease. In other words, death rates for light to moderate drinkers are slightly below those for complete abstainers, although these rates increase with higher drinking levels. The exact shape of the curve varies in different populations but the "J" shape has been most pronounced in populations at higher risk for CHD. Where this risk of heart disease is low — in less-developed countries or among people under 45 years of age — the "J" is less pronounced or disappears completely, and the relationship between drinking and death rates is simply linear — as drinking levels increase, so do death rates. Evidence for a protective effect of alcohol on cardiovascular risks also comes from recent studies of people with an eye condition known as branch retinal vein occlusion. This condition is suspected of being related to arterial disease and there is evidence for reduced risk with increased alcohol use (Eye Disease Case-control Study Group, 1993).

There are several possible explanations for the finding that, in some samples, light to moderate drinkers have lower mortality rates than complete abstainers. Thus, the findings may result from a failure to adequately control for other factors such as family history, diet, smoking, exercise and use of drugs other than alcohol. It is also possible abstainers have higher CHD rates because some are former heavy drinkers or have other health problems that motivate them to be abstinent. It is also possible that light to moderate drinking "protects" the drinker against CHD. This is the view promoted by the media and is much favored by the alcohol industry.

Much has been made of data that show lower-than-expected rates of heart disease in France where there is a long tradition of high fat and high alcohol consumption. Some researchers have argued that the French data show that alcohol, and especially red wine, counteracts the effects of fat in the diet. This view has been challenged on two accounts: first, on the grounds that the intake of fat in France is a relatively recent phenomena that may yet lead to higher rates of heart disease (Nestle, 1994); second, on the grounds that the French drink less milk (Popham, Schmidt and Israel, 1983). Milk contains proteins known to contribute to fatty deposits in the coronary arteries, making high consumers vulnerable to heart disease.

Biological explanations for the apparent relationship between low levels of drinking and reduced heart disease include alcohol's effects on cholesterol levels, blood clots, insulin sensitivity and estrogen levels among post-menopausal women (Rankin, 1994). Some research has suggested that resveratrol, an antifungal agent in grape skins, may benefit wine drinkers and especially those who drink red wine (Nestle, 1994). However, there are many unanswered questions and some researchers remain sceptical of the whole notion that low to moderate drinkers have a reduced risk of heart disease. It has also been pointed out that the alleged benefits may accrue at very low levels of drinking (perhaps as low as one drink every other day) and then only for middle-aged men (Anderson, 1994). Further, there are many people, including those at risk for some types of heart diseases, for whom low levels of drinking are contraindicated. Also, apart from drinking, there are many things that people concerned about heart disease can do to reduce the risks, including not smoking, eating low-fat foods and getting an appropriate amount of exercise. As a result, many experts tend to agree that there is no good reason for abstainers and very light drinkers to increase their alcohol use in order to reduce their risk of heart disease. To do so, without making other lifestyle changes, may have little benefit and may increase the risk of other health problems.

How Do Canadians Respond to Alcohol Problems?

In the next chapter, we explore Canadians' continuing support for a wide range of government measures and institutions regulating alcohol and its availability. Through governments, and in their hearts, Canadians also support treatment programs for people with alcohol problems (see Chapter 12). But how do they respond to family members, friends, co-workers and acquaintances with alcohol problems? A recent survey in Ontario throws light on this question (Bondy, 1994).

This survey involved telephone interviews with over 900 people chosen to represent the adult population of the province. Among other things, they were asked if they had said anything to a friend or family member who had a drinking problem and, if so, had they recommended getting help. Fifty-eight per cent of those interviewed said that a family member or a friend had an alcohol problem at some time and, for 25 per cent of respondents, this problem was during the previous 12 months. Of those with a friend or relative with a problem in the past year, 34 per cent had said something about this problem to the person concerned and 14 per cent had suggested getting help. Men and women were equally likely to report making a comment, but those aged 18-34 were more likely to report making comments than their elders.

Most of those making comments to a friend or family member with a drinking problem reported that they had thought about saying something before, but had not done so. They also indicated that alcohol-related family problems led them to comment when they did. Of those who reported that they recommended a friend or family member get help for drinking problems, 58 per cent recommended Alcoholics Anonymous while 29 per cent recommended a doctor or nurse. One in ten recommended detoxification centres while only eight per cent recommended a specialized alcohol-treatment agency. These results seem to indicate that Canadians are sometimes reluctant to talk about alcohol problems to those who experience them, and that when they do talk about these problems, getting help from specialized treatment services are not high on the list of options considered.

This survey also included questions about steps respondents had taken to prevent various alcohol problems. The majority, especially among women, responded affirmatively to questions about serving food and non-alcoholic beverages to guests and making sure guests did not drink too much, or drink and drive (West, Dick et al., 1994).

Summary

Alcohol is directly or indirectly involved in many health, social, and legal problems in Canada. Among the more serious problems associated with the intoxicating effects of alcohol are traffic accidents, violence, poisoning, adverse reactions with other drugs, and fetal damage. The adverse consequences of long-term heavy drinking include alcohol dependence, liver disorders, brain damage, certain types of cancer, heart damage, and gastric and other medical disorders. Long-term heavy use is also frequently associated with psychological, social, and legal difficulties. Some population studies have suggested that moderate alcohol consumption reduces the risk of heart disease and this is supported by biological research. However, many experts doubt that the benefits of drinking are significant and do not recommend drinking as a health measure.

Wagon carrying a petition roll for "a dry Ontario" during a Prohibition parade (Toronto, March 8, 1916)
John Boyd/Public Archives of Canada(PA-072527)

The Citizen's Committee of One Hundred arrives a Queen's Park to present the petition calling for Prohibition (Toronto, March 8, 1916)
John Boyd/Public Archives of Canada(PA-072520)

A half-mile of barmen carry a banner on College Street (Toronto, March 8, 1916)
John Boyd/Public Archives of Canada(PA-072525)

Barmen march down Yonge Street (Toronto, March 8, 1916).
John Boyd/Public Archives of Canada(PA-072524)

Beer issued to the 92nd Highlanders aboard the Empress of Britain (May 1916).
Public Archives of Canada(PA-022720)

Street arrest during Prohibition (Toronto, September 1916).
John Boyd/Public Archives of Canada(PA-069901)

Carrying a keg home during Prohibition (Toronto, September 1916).
John Boyd/Public Archives of Canada(PA-069965)

Chapter 8

Alcohol Production, Taxation and Controls In Canada

Most alcoholic beverages consumed in Canada are domestically produced although imports have become far more important since the early 1990s because of the GATT (General Agreement on Tariffs and Trade) and NAFTA (North American Free Trade Agreement) treaties. Now there is a substantial export trade in Canadian beer and spirits, especially to the United States. In this chapter we describe the extent of alcohol production in Canada, the way it is taxed, and efforts made by governments to control its sale. We will discuss international efforts to control alcoholic beverages. We will also consider issues concerning privatizing the sale of alcoholic beverages — a major change that could have important implications for governments in Canada. Evidence of the impact of various alcohol-control measures on consumption and related problems will be explored in Chapter 9.

Alcohol Production

Beer Production

The brewing industry in Canada is a relatively large one that has traditionally been dominated by a few giant corporations. The trend up until the late 1980s has been for small local or provincial breweries to disappear, usually swallowed by one of the larger companies. However, in the past ten years or so many small breweries and micro breweries have been established. Currently, there are two large companies controlling most of the market, five regional breweries and 56 micro breweries. There are also about 300 "U Brew" facilities in Ontario and British Columbia. The large companies are Molson and Labatt's, and both market beverages in Canada and worldwide. Probably beer companies are worried about the long-term prospects for beer consumption, as both Labatt's and Molson have acquired wineries. These two companies now have a wide range of business interests, and each owns a professional sports team.

Imports of beer from the United States have increased so they are far more important than before 1975 (Brazeau and Burr, 1992). Almost all beer sold in Canada is technically Canadian made; only 3.3 per cent of beer drunk in Canada is imported. However, sales of licensed international brands made in Canada, such as Miller or Budweiser, account for 10 to 15 per cent of sales — and that percentage is growing. Beer imported into Canada comes mainly from the United States, with smaller amounts from the Netherlands, Denmark, other European countries and Australia. Virtually all beer exported from Canada goes to the United States or the U.S. Virgin Islands, although small amounts go to many other countries.

Many changes have recently occurred in beer production in Canada. Low-alcohol beers and "near beers" were introduced in the mid-1980s. These typically contain 2.4 per cent and less than one per cent alcohol respectively. Tastes have shifted to such beers so that they now represent about 13 per cent of the market. Near beers are sold in grocery stores and hence there has been some concern that children could buy them and get drunk on them. However, that seems unlikely unless very large amounts were taken.

High-alcohol beers have also been introduced in Canada with as much as 9.5 per cent alcohol content, compared to about five per cent for regular beer. This has led to criticism of the breweries because of the higher risks of alcohol-related accidents and drunkenness. As a result of this criticism some higher-alcohol beers have been withdrawn.

Beer production has also changed because many smaller producers have come on the scene. "U Brew" operations have proliferated in B.C. and Ontario. These stores — where you can help experts make beer for you — are in fact small breweries. Until recently, such operations were not taxed and hence were very cheap, but a low tax rate has now been imposed. There are also micro breweries, which brew small amounts of specialized beers, and brew pubs, which can brew beer for sale in their own pubs. As yet, these new facilities have gathered only a few market-share percentage points. However, they could be more important in the future.

Spirits Production

Spirits production in Canada is also a large-scale activity dominated by a few firms. Most of the distilling companies are controlled by several large firms: Seagram, Hiram Walker, Gilbeys and Carringtons. All operate a variety of distilleries and market spirits under several different brand names. As with the beer industry, the tendency has been for the giants to take over smaller firms and to have a large export market. For the two largest firms, the Canadian market is only a small part of their worldwide sales. The Seagram Company is one of the largest distillers in the world and sells its products in nearly every country where there are drinkers.

As with beer, most spirits consumed in Canada is produced here, but the market share of imported spirits is changing very little. There has been a long-term movement away from whisky, or "brown" goods, and gin towards vodka and white rum. Most spirits contain about 40 per cent alcohol but liqueurs vary between 18 and 24 per cent. "Coolers" were introduced in the mid-1980s and contain about 5 per cent alcohol. Another effort to recoup failing market share has been the "light" distilled spirits beverages introduced in 1990 but it is not yet clear how popular they are.

Wine Production

Canadian wine production is structured differently from spirits and beer production. The firms are small, have no export market to speak of, and have trouble even controlling the domestic market. Almost 70 per cent of the wine sold in Canada is imported and that proportion is increasing rapidly.

There are about 35 wineries in Canada. All are small by industry standards and most sell their products in only one or two provinces and rarely export much wine. The four largest companies are Andres, Brights, Chateau-Gai, and Jordan. The largest wineries sell about 80 per cent of domestic wine consumed (Brazeau and Burr, 1992). However, the number of wineries is increasing in Canada. Several smaller wineries make a limited range of high-cost products from newly developed varietal grapes. There are also a number of boutique wineries with very small specialized products, as well as some wine bars and make-your-own wineries. The usual tendency for domination by multinational giants has not come to the Canadian wine industry. Of course, the small size of domestic production and consumption probably makes Canadian viniculture an unattractive target for the giants.

Prices and Taxation of Alcoholic Beverages

Just how the prices of alcoholic beverages are set is not easy to understand. Technically speaking, prices are set by the various alcohol monopolies and commissions and the federal Government. However, there are many influences operating. Producers, of course, want a return that covers costs and allows for profits. Imported beverages will show some price fluctuations because the governments exporting them demand more profits or because of changes in the international currency market. An international review of alcohol taxation is available in the review by Brazeau and Burr (1992).

The major cost determinants are, of course, the taxation and revenue-generating policies of the various governments involved. In Canada, taxes on beer account for about 53 per cent of the consumer price of beer, 69 per cent of the price for domestic wine and 82 per cent of the price for spirits (Brazeau and Burr, 1992). The rationale for the difference is not clear; possibly governments favor low-alcohol beverages on the expectation that their use is less dangerous to individuals and public safety. The evidence for that belief is, however, poor; health and safety are more related to how much alcohol is drunk than to the particular beverage type (see Chapter 10 for a review of these issues).

The rules for the taxation of alcoholic beverages sold in Canada are unbelievably complex. There are several types of taxes at both the provincial and federal levels plus provincial price "markups." Excise levies at the federal level vary considerably depending on the type of beverage (much higher for spirits than for beer). There are also sales taxes at both levels. It is thus very difficult to arrive at exact amounts of net tax per unit volume. However, it has been estimated (Brazeau and Burr, 1992) that the total tax per litre of absolute alcohol is about $32.19 for beer, $29.81 for table wine, $19.88 for fortified wine and $51.14 for spirits. The cheapest alcohol for the consumer usually turns out to be either beer or fortified wine. Canada is one of the countries with the largest tax burdens on alcoholic beverages. Surprisingly for products taxed so highly, alcohol contributes only 2.7 per cent of all taxes collected. Canada is, however, one of the top five countries in terms of taxes collected on alcohol (Brazeau and Burr, 1992).

Some Benefits of Alcohol Consumption and Production

Today, the production and distribution of alcohol represent significant economic activities in Canada and alcohol sales taxes alone generate about $4 billion in federal and provincial government revenues. For provincial governments, alcohol sales taxes account for 1.3 to 2.8 per cent of total revenue (2.1 per cent, on average). In addition, Canadian governments benefit from taxes paid by workers involved in alcohol production and trade, from workers in associated industries (hotels, tourism, advertising, bottle-makers, etc.) and from corporate taxes paid to government by the alcohol industry. Total government revenues associated with alcohol production and distribution are thus very large, although the exact figures are unavailable.

The latest available figures show that the alcohol beverage industries employ about 19,000 people, whose wages and salaries amount to $921 million (Single, Williams et al., 1994). In addition, many other Canadians are primarily employed in distributing alcoholic drinks (e.g., waiters and waitresses and bartenders, employees of provincial liquor control boards and of Brewers' Retail). Many other Canadians also work with or for organizations that produce or sell alcohol, including farmers, bottle manufacturers, architects and builders, people working in advertising, musicians, etc. However, the extent to which alcohol production and sales impact these and other groups has not been researched.

Cynics may also want to include those who make a career in the "alcoholism business" (e.g., treatment personnel, researchers and administrators of alcohol programs) among the beneficiaries of Canadian drinking practices. To these we might add the many police officers, court officials, social workers, and family counsellors dealing with alcohol problems. Certainly a sudden drying up of alcohol could pose employment problems for those involved in the processing and care of problem drinkers and charges of self-serving "scare mongering" have sometimes been levelled at care givers and others on the "problem" side. However, this is clearly unfair because resources given over to the treatment of people with alcohol problems represent only a fraction of those directly and indirectly generated from alcohol sales. For example, in Ontario the provincial government receives over $1 billion from alcohol sales taxes and licences but spends only one-tenth of this amount on specialized addiction treatment services (Ellis and Rush, 1993).

Enthusiasm for the economic benefits of alcohol production and sales should be tempered by an awareness that much of the alcohol is consumed by heavy drinkers. This is known from studies of buying and drinking habits and from what we otherwise know about the distribution of alcohol consumption (see Chapter 4). In common with other alcohol-using countries, Canada has a high proportion of people who use relatively little alcohol, a smaller proportion of "moderate" drinkers, and relatively few "heavy" drinkers. In the late-1960s it was estimated that if "heavy" use is defined as drinking in excess of 6 drinks a day, then the seven per cent of Canadians who typically drank at or above this level accounted for almost 50 per cent of all the alcohol consumed in the country (Hyland and Scott, 1969).

Smuggling and Illegal Importation of Alcohol

With a long undefended border with the United States we should expect alcohol smuggling to be a problem. Many U.S. border states have private enterprise systems for alcohol sales and much lower prices for some beverages — especially if import duties are not paid. Alcohol that is smuggled is lost to the government accounting systems, is untaxed, and is not reflected in per capita consumption figures. Sometimes smuggling operations are very large and involve illegally made beverages. For example, a single task force in the Akwesasne Mohawk territory in Quebec seized 10,000 1.75 litre bottles of illicit spirits in May 1994 (*Globe and Mail*, June 29, 1994). These bottles, which can be bought in the U.S. for $2, sell for $30 to $35 in Canada. Often they are sold to restaurants and bars where they are used in mixed drinks without the customer knowing. It has been estimated that $1 billion a year is lost in tax revenues solely from illegal smuggling. A recent study by the Liquor Control Board of Ontario estimated that 15 per cent of the province's $5.4 billion alcohol market is lost to smuggling (Churley, 1994). Of course, these estimates are based on many assumptions but the smuggling problem has certainly increased in recent years.

Another way alcohol gets into Canada is by travellers from the U.S. or Canada returning with legal or illegal purchases. The number of trips by Canadians to the United States increased greatly during the late 1980s as inflation and taxation levels increased in Canada. It has been estimated that about half of these three million travellers bought alcohol back to Canada — again outside of our accounting system. Good estimates of the amount of alcohol brought into Canada by travellers are hard to find.

Alcohol Controls and Alcohol Policies

Alcohol controls have been a controversial and rapidly changing aspect of public policy in Canada. As alcohol policies become liberalized, they also become much less important politically. However, public controversy still surrounds questions about alcohol advertising, whether beer and wine should be sold in grocery stores, and whether provincial retail monopolies should be privatized. Canadian alcohol-control policies are complex, with laws and regulations varying by province and changing over time. This section describes the background of controls in Canada, their current status and rationale, and the controversy surrounding them. Generally alcohol controls are set with community standards, economic matters and sometimes public safety in mind. Whether a particular change in controls would affect public health seems to be rarely considered in Canada.

Federal-Provincial Jurisdiction over Alcohol Sales

The constitution in Canada is currently a combination of the new Canadian constitution and remnants of the *British North America (BNA) Act*. Together they define the powers of the various levels of government and of the courts. *The BNA Act* of 1867 set out the most important rules for the establishment of institutions and the use of political power in the country, and so far these have not been changed under the new Constitution. *The BNA Act* delineates the powers held by the federal and provincial governments. Alcohol-control policies, alcohol sales, and alcohol distribution are not specifically mentioned in the Act. It states that "in each province the legislature may exclusively make laws in relation to . . . Shop, Saloon, Tavern, Auctioneer, and other licenses . . . [and] generally all matters of a merely local or private nature in the province." Also, trade within provinces is regulated by the provinces themselves. These provisions are taken to locate most aspects of alcohol control in the provincial powers. Neither the framers of the *BNA Act* nor the new Constitution paid much attention to alcohol policy or controls. In practice, alcohol-control policies, such as age of access to alcohol, numbers and types of outlets, distribution networks, types of beverages and hours of sale are governed by the provinces.

The federal government does, however, control some aspects of how alcoholic beverages are marketed. For example, the federal government controls customs and importation. It also taxes alcoholic beverages in all provinces. Because broadcasting is federally controlled, the types of alcohol advertisements on radio and television are limited by federal regulations. Alcohol sales on airplanes and in airports and to aboriginal people are also federally regulated. A mixture of both federal and provincial regulations govern packaging, labelling and the purity of the ingredients of alcoholic beverages. However, the alcohol-control system is chiefly a provincial one and is seen to be so by citizens and politicians.

Because alcohol-control policies are mostly provincial, there is no one unified system of controls. There are ten provincial systems and one each for Yukon and the Northwest Territories, with many local variations on Indian reserves and in Inuit communities. There is little effort to harmonize these systems and none to provide central control or management. However, while each system is operated independently, there are many similarities.

Current Status of Major Alcohol Controls

The Monopoly System

In Canada, alcoholic beverages are made available chiefly through government monopoly systems (see Brazeau and Burr, 1992 for a review). Since the repeal of Prohibition, each province has operated a liquor control board, commission, or corporation. Since the 1960s these provincial monopolies have seen a gradual relaxation of their control function and a growing emphasis on "customer service" and revenue generation. More recently, large-scale privatization has threatened the future of the government monopolies. In 1994, the Alberta government privatized its provincial retail system. Similar measures have been under consideration in Quebec since the early 1990s. The proposed privatization of the Liquor Control Board of Ontario was part of the Ontario government's successful 1995 election platform.

The system of provincial liquor boards and commissions has virtually total control over which alcoholic beverages can be sold, how and when they are sold, and to whom. They also set prices and taxes for alcoholic beverages (along with the federal government) and govern the number of outlets of all types. This system differs considerably from that in most U. S. states and many European countries, where prices and availability are set more by market forces rather than government regulation. Nonetheless, the Canadian alcohol monopolies typically are marketing agencies with a profit motive. Their purposes are to create an orderly market for alcoholic beverages, make them available at reasonable cost to consumers, and generate profits for governments.

In the past few years many questions have been raised about the effectiveness of alcohol monopolies. Some business interests have claimed that private enterprise could run the stores more efficiently and that governments should not be in the "liquor business" at all. Government monopolies, however, are common where (i) a stable market is needed, limited by demand not supply; (ii) there is a desire to eliminate the profit motive, (e.g., electricity or water supply), and (iii) where there are health or public safety concerns. Alcohol monopolies seem to fit all of these requirements and hence make good public policy sense.

In the debate over the future of Canada's alcohol monopolies, it is interesting to compare the social policy effectiveness of monopoly and private enterprise systems. In Canada we cannot compare monopoly and licence systems for alcohol but American studies can. At present, there are 14 control or monopoly states and 44 licence or private enterprise states in the U.S. and several comparisons have been made. Several studies (Surion, 1966; Nelson, 1990) have shown that there are fewer stores per capita in monopoly states. Nelson's study (1990) showed that there were four times as many stores in licence states. He also found that per capita sales of beer and spirits were slightly higher in licence states but about 15 per cent higher for wine and spirits. However, prices for all these beverages were the same in licence and monopoly states. Unfortunately, Nelson's data could not include figures for home production, nor legal or illegal importation by individuals. It appears, however, that in monopoly states alcohol costs about the same as in licence states but consumption is lower, probably because of the lower number of stores. However, other factors such as temperance sentiment and economic factors such as lower incomes may also be important. Of course, monopoly states generate more taxes per bottle for governments because all profits go there and are not shared with private store owners.

Sales Outlets

The various government alcohol monopolies exert strong control over the distribution of alcoholic beverages. None of the monopolies actually produce alcoholic beverages of their own. However, some import them in bulk and bottle them with their own labels. In fact, all production of alcohol in Canada is in the hands of private companies.

The relative proportions of different types of outlets vary somewhat from one province to another within a rather narrow range. In no province do monopolies operate restaurants or bars, and their retail activities are limited to "package stores" of different types. The most usual situation is that beer, spirits, and wine are sold through the monopoly's own stores, with beer also available through brewery stores, licensed vendors, or grocery stores. Ontario has a small number of privately owned wine stores operated by wineries and a large number of "agency" stores located in grocery stores in sparsely populated areas.

A few provinces allow the sale of alcoholic beverages in grocery stores. Quebec allows the sale of beer and wine (not spirits) in grocery stores, as well as in the liquor corporation's own stores. Ontario does not technically allow the sale of wine in grocery stores, but a few large stores have small booths for wine sales operated by the wineries themselves; wine purchases must be paid for separately and cannot be part of the food sales. Beer is sold in food stores in Newfoundland. In February 1994, Alberta allowed sales of all alcoholic beverages in independent grocery stores as part of its move to privatize alcohol retailing. No other provinces allow the sale of alcoholic beverages in food stores except for cooking wine and Chinese wines, which are heavily seasoned and not suitable for drinking.

Packaged beer is made available in Canada in a number of ways. Typically, imported beer is available only through monopoly stores. Local beer is sold only through such stores in the Maritime Provinces. However, in Ontario beer is sold not only through monopoly stores but also at breweries and at stores owned by the brewery consortium. Western provinces have sales through board stores but also through hotels with off-premise licences. The Canadian drinking public generally has far better access to beer than to wine or spirits.

On-premise consumption is allowed at a variety of provincially licensed establishments:

- *Taverns, public houses, or beer parlors* typically sell only beer (with or without food). In some provinces they sell wine as well. Recently western provinces have allowed the sale of wine and spirits in these previously "beer-only" places.

- *Restaurants* can sell alcoholic beverages only with meals. The majority can sell all three types of beverages, but some are limited to beer and wine.

- *Lounges* sell all alcoholic beverages without any food requirement.

- *Private clubs* may have the right to sell any or all beverages on a permanent basis.

- *Sports areas, stadia* and *ball parks* in some provinces (e.g., Ontario and Quebec) are allowed to sell beer, and sometimes wine, at a variety of sporting events, such as professional hockey, football, baseball and basketball games and tennis and golf tournaments.

- *Churches, clubs,* or *other institutions* may get licences to sell beverages, usually only wine or beer, for specific events such as weddings or parties.

- *Individuals* may get special-occasion permits for weddings, large parties or socio-cultural events. These may be "sale" or "no-sale" events.

Local option laws in most provinces allow voting on the introduction of different types of outlets. In most parts of Canada such votes have been won by liberal forces, and consequently alcoholic beverages are available nearly everywhere. However, some anomalies exist. For example, local option has kept alcoholic beverages from being sold in any form in a small area of Toronto. Several communities in the Northwest Territories have also voted to have Prohibition introduced (Smart, 1979a).

Visitors to Canada usually find alcoholic beverages less available than in most European countries or most states in the United States and about the same as in most Scandinavian countries. Wide-open, round-the-clock availability, such as is found in California or Alaska, does not exist legally anywhere in Canada.

Hours of Sale

The hours of sale in Canada vary considerably and change frequently. Most monopoly package stores are open for business for 8 to 12 hours per day with no sales on Sundays or holidays. On-premise consumption is typically allowed from noon until between midnight and 3 a.m., with a break for the dinner-hour in some provinces. Formerly, on-premise consumption on Sundays was not generally allowed, but now some provinces allow Sunday sales of wine and beer and on-premise drinking whether a meal is also ordered or not. Quebecers have beer and wine and Newfoundlanders beer available in small grocery stores at variable hours, from nine in the morning until late at night. Manitoba, Saskatchewan, Alberta and British Columbia have beer (also wine and cider in B.C.) available from hotels licensed to sell for off-premise consumption.

Advertising of Alcoholic Beverages

There is a wide range of laws and regulations affecting alcohol beverage advertising in Canada. Prince Edward Island permits virtually no advertising of any kind. Saskatchewan recently rescinded a ban on alcohol advertising. Ontario and Quebec are the most liberal about advertising, but neither allows large exterior billboards. Ontario allows advertising of alcohol on subway and bus stop posters. Canadian provinces have far less alcohol advertising than do European countries and most U. S. states.

The advertising of alcoholic beverages is limited by both federal and provincial regulations. Until 1995, federal regulations prohibited the advertising of spirits on radio or television, while magazine and newspaper advertising of spirits is allowed in several provinces. In most provinces beer and wine can be advertised in all media, but wine is rarely advertised on radio and television, except at Christmas and New Year, because of the high costs relative to its share of the market.

Even where advertising is allowed, there are numerous regulations as to types of advertisements allowed, who can be in them (no children), what they can portray (fun, but no problem-solving effect of alcohol) and where they can be presented (e.g., not on children's television programs).

While efforts to control alcohol advertising have attracted much attention, these efforts have also been the focus of a good deal of criticism. Studies by Health and Welfare Canada have indicated that the guidelines of the Canadian Radio and Television Commission (CRTC) have been often breached. For example, alcohol ads often appear on programs watched by children, e.g., hockey games. They often contain actors who look very young (although they are really over the drinking age), some beer ads appear to be specifically directed at young people (using rock music, youthful themes, etc.) and in some cases celebrities (former hockey players, movie stars, etc.) have advertised alcoholic beverages. These rules are often vague and difficult to enforce and some (such as the ban on celebrities in alcohol advertising) would not stand a court challenge.

As argued at length in Chapter 9, advertising restrictions may not be a very important factor in alcohol consumption. Advertising bans have not been very effective in Canada. Manitoba banned beer advertising in electronic media in the 1970s as did British Columbia, but there seems to have been no effect on sales (Ogborne and Smart, 1980). Saskatchewan lifted a 50-year advertising restriction for beer, wine and spirits in 1988 but there seemed to be no effect on sales (Makowsky and Whitehead, 1991).

For many years the CRTC had rules against the advertising of spirits on Canadian electronic media. However, that has had to change. Spirits manufacturers argued in 1991 that they were unfairly treated and that under the Constitution they should have the same right to advertise as beer and wine interests. The CRTC held hearings and received many briefs from interested parties. Most of those such as the Addiction Research Foundation argued against spirits advertising as it may increase consumption. Some groups such as the Canadian Centre on Substance Abuse said that adding spirits would make little change in overall alcohol consumption. In 1995, a federal Court decided that spirits could be advertised on the electronic media. The decision seems to have been based largely on constitutional issues of fairness in advertising and freedom of speech as well as a feeling that spirits advertising would not constitute an overriding health concern. However, as of mid-1996, spirits manufacturers had yet to advertise on television. It is expensive to do so and the small market for most spirit-based beverages may make it impractical for many distillers.

Legal Drinking Age

The legal drinking age for alcoholic beverages is 18 in some provinces (Prince Edward Island, Quebec, Manitoba, Alberta) and 19 in the rest. All provinces lowered the legal drinking age from 20 or 21 in the period between 1970 and 1975. The change was part of a general movement towards lowering and standardizing the age of majority. Findings by several provinces that these age reductions led to increased drinking and drinking problems among young people soon led to debate about the age law in many provincial legislatures (Smart and Goodstadt, 1977). Consequently, Saskatchewan and Ontario increased their drinking ages from 18 to 19 in the late 1970s. There is no differentiation as to what beverages can be sold at the legal age.

International Influences on Alcohol Controls

Traditionally, provinces have established alcohol policies without much reference to other provinces, the federal government, or what was going on in other countries. In the late 1980s this began to change radically. In 1988, Canada agreed with the European Community (Dunning, 1993) to reduce store prices for beer and brandy from the EC and to give wines and spirits from the EC equal access to the Canadian market — chiefly by offering more shelf space. In 1989 the *Free Trade Agreement* (FTA) with the United States went into effect. Among its provisions, the FTA required Canada to reduce price and other barriers to the sale of U.S. wine and spirits. This meant that preferential treatment of Canadian wines disappeared and that cheaper California wines would be more readily available. Because beer was excluded from the FTA, U.S. breweries filed a complaint under the General Agreement on Tariffs and Trade (GATT), winning greater access to beer stores in Canada at reduced prices. All of these changes will serve to lower prices and make a wider variety of products available. We might expect some overall increase in alcohol consumption, however, imported products may just take some market share from domestic producers without affecting per capita consumption.

Some History and Current Policy Issues

Alcohol-control policies have been both public and political issues in Canada. In the early 1900s, politicians could not be elected in many areas without a clear temperance plank in their platform. Temperance issues were frequently debated in legislatures. As we saw in Chapter 2, most provinces voted for Prohibition in the early part of the century. However, the temperance movement has since lost its political base and now holds little influence over political life in Canada.

The present level of availability of alcohol in Canada is fairly recent and follows a change in public attitudes about alcohol. Prior to the conclusion of the Second World War, many provinces had no on-premise consumption of spirits, no bars or lounges, and few restaurants where alcohol could be purchased. Beer parlors and hotels where only beer was sold, frequently only to men (although some had "women's sections"), were common. The postwar period brought many social and attitudinal changes and politicians saw a clear need to bring liquor laws into step with more liberal public attitudes. The period from 1948 to the present is one in which physical availability of alcohol to Canadians greatly increased. The number of changes is too great to document fully here, but the list for Ontario is similar to that for most provinces, and includes the following:

1948

Lounges and bars are permitted to sell alcohol for the first time.

The 1950s

Local option votes could be held before a licence was issued rather than afterward.

Drinking is permitted in a trailers or tents for the first time.

Beer parlors are allowed to sell food.

Local option votes are allowed in areas with populations of 50,000 or over.

Liquor-permit books are replaced by permit cards. Previously anyone buying for off-premise consumption had to present a permit book in which each purchase was recorded.

The 1960s

Allowance is made for carrying bottles containing alcohol from the store to a residence. Previously it was illegal to have a bottle anywhere but in a residence, including between the store and the residence. Alcoholic beverages could be given as a gift.

Liquor-permit cards are discontinued altogether.

Private clubs are allowed to have licences in otherwise dry areas.

Hours of sale for on-premise consumption are extended to up to 12 hours per day.

Hotel room service is allowed to sell alcoholic beverages.

Motels and summer resorts are allowed to apply for licences for the first time.

Wine could be sold that contained fruits other than grapes.

Licences could no longer be cancelled without a hearing.

The right to appeal interdiction proceedings is introduced. The interdiction list specified individuals prohibited from purchasing alcohol.

Hotels in small towns could apply for licences.

The sale of alcoholic beverages on airplanes is allowed.

The sale of alcoholic beverages in live theatres is allowed.

Public houses are permitted to remain open between 6:30 p.m. and 8:00 p.m.

Drinking on patios and in backyards of private homes is allowed.

People in hotel rooms who are not actual residents are allowed to drink.

Self-service liquor stores are introduced. Formerly all stores had been clerk-service stores. Within a few years most stores are converted to self service.

The 1970s

Unsegregated seating of men and women in licensed establishments is allowed. Previously, a man could enter licensed premises that served women only if he was accompanied by a woman.

Licences could be granted to resort centres in dry areas of the province.

Rules requiring dining-lounge patrons to order a meal in order to have a drink are abolished.

Drinking is allowed with meals on Sunday in licensed restaurants.

Permit requirements are relaxed for special events, such as fairs, festivals, winter carnivals, Oktoberfest, etc.

40 oz. bottles of spirits are permitted. Previously 26 oz. was the largest bottle available.

Retail stores are allowed to stay open until midnight.

Duty-free stores at airports are allowed to sell alcoholic beverages.

Licensed establishments are allowed to open after polls had closed on voting days.

Licences could be issued to new premises without a waiting period. Previously a restaurant would have to be open six months before being licensed.

The legal drinking age was lowered from 21 to 18 (in 1979 it was raised to 19).

Definitions of recreational facilities and theatres eligible for licences are broadened.

Licences are allowed for canteens on campuses of colleges and universities.

Parents are allowed to serve alcohol to minors in their own homes.

Bottles with the seal broken (i.e., partly consumed) could be transported to any destination.

The 1980s
Wine kiosks located in grocery stores are permitted.

"Happy hours" introduced in 1982, and banned in 1984.

The 1990s
Drinking is allowed at professional casinos and during gambling activities

Brew pubs, U-Brews and wine pubs are allowed

Beer sales are permitted in certain large arenas and stadia.

Home delivery of liquor by specially licensed agents is permitted.

Drinking is allowed in parked boats and motor vehicles if they have eating and sleeping facilities.

Out-of-province and imported beer gain equal access to Ontario beer and liquor stores.

Sunday sales of beer at breweries and wine at wineries and wine stores is allowed.

Purchasing alcoholic beverages on credit cards is permitted.

Of course, none of these changes is significant in isolation. However, their combined impact was to change the temperance-oriented, mostly dry province of Ontario into one where alcohol is cheap and available in most social and cultural situations with no bureaucratic interference. Similar changes were enacted throughout the country, probably at somewhat different times and in a different sequence, but the effect was very similar.

The liberalization of alcohol-control laws abated slightly in Ontario in the late 1970s and early 1980s, when few new laws were enacted to make alcohol more available. However, changes in their administration led to the granting of licences on university campuses and allowed the sale of beer in certain ball parks. Also, in 1982 "happy hours" were allowed in Ontario — that is, hours when alcohol could be made cheaper or sold on a "two-for-the-price-of-one" basis. However, late in 1984, "happy hours" were once again banned, and now drinks in bars and taverns must cost the same at all hours of the day. Surprisingly, the elimination of happy hours had no effect on drinking or drinking and driving (Smart and Adlaf, 1986).

Since 1980 several communities in Northern Canada have introduced Prohibition, while others are considering it. Also, some provinces have limits on alcohol advertising for electronic media, and some provinces (e.g., Ontario and Saskatchewan) have strengthened advertising codes to limit "lifestyle" advertisements. Several provinces have debated the issue of introducing wine and beer into small grocery stores and have rejected it for the time being. Local merchants and tourist agencies greatly favor such a change, although large supermarket chains do not. In Manitoba and Ontario, these plans have been put on hold although wine can be sold in special kiosks in Ontario.

Public Opinion about Alcohol Controls

Many surveys in Canada have asked the public their opinions about alcohol policies. The largest national survey of this type was the *1989 National Alcohol and Other Drugs Survey* (NADS). In general, the results show that most Canadians are conservative on such issues. For example the NADS found that a majority want:

(i) taxes on alcohol to increase (27 per cent) or stay the same (46 per cent)

(ii) beer and liquor store hours to stay the same (70 per cent) or decrease (17 per cent)

(iii) alcohol not to be sold in corner stores (74 per cent).

On the other hand, most Canadians want warning labels on alcoholic beverages (74 per cent), a ban on alcohol ads on television (51 per cent, with 6 per cent undecided), and increased efforts to prevent drunks being served (81 per cent). In general, there seems little support for more liberalization of alcohol controls. However, most Canadians want more government educational efforts to reduce the risks from drinking.

Privatization of Alcohol Sales: A Coming Flood or Just a Trickle?

After wine sales were allowed in Quebec grocery stores in 1979 many people believed that full privatization would come soon in Quebec and other provinces. However, it was not until 1994 that any province fully privatized its alcohol-retailing system. In 1984, the Ontario Liberal party included partial privatization as part of its election platform, but legislation to introduce beer and wine into grocery stores was defeated under the subsequent Liberal minority government. Of course, we have seen some creeping privatization of wine and beer sales in Ontario and other provinces. However, by 1995 no province other than Alberta had fully privatized alcohol sales. It is too early to tell what the effects of privatization in Alberta have been. However, there have been some complaints about the prices being too high. Also, at least one grocery chain — Safeway — was reluctant to get into such sales, presumably because profits may be too low and the field may already be saturated (*Edmonton Sun*, June 2, 1994).

Whether privatization of alcohol sales is seen as a good policy option depends on many factors. On the positive side, some governments would like to get out of running retail businesses and just collect the profits. They receive criticism for alcohol policy, high prices and poor service as well as occasional strikes that cut off supplies to consumers. But could governments maintain the same level of profits if they did not operate the stores? We have seen that monopoly systems have fewer stores than licence systems, which must create efficiencies in staffing, transportation and storage costs. In order for many private entrepreneurs to profit from liquor and wine stores prices would likely have to rise. If prices were to rise very much people might buy less, reducing profits for everyone. Another less likely possibility is that governments might accept a lower level of profit in order to let private entrepreneurs enter the field. We should remember that the monopoly system generates the most profits for governments and if they give up distribution some profits will be threatened.

We don't yet have very good evidence about the effects of privatization in Canada. In Quebec the privatization of wine sales had no effect on total alcohol or wine consumption. However, prices of wine sold in grocery stores were about nine per cent higher than in government stores (Smart, 1986), which probably held consumption down. Privatization was recently introduced into Iowa, where some before- and after-studies have shown small increases in consumption and others have not (Holder, 1993). What conclusion is made about Iowa probably depends on what analysis is employed.

Privatization of alcohol sales in Alberta is relatively new. The outcome there may influence the future of alcohol monopolies in other provinces. Preliminary evidence from Alberta shows that following privatization the number of stores increased greatly, so that such stores are probably closer to many people's homes than in the past. However, the selection of beverages has not increased and prices have risen by as much as 15 per cent, so the experiment is not a success for the consumer. However, more time will be needed to be sure of the effects of privatization in Alberta.

In the early and mid-1990s, alcohol availability seems to have increased again to new levels. Some of these increases are due to international pressures and some to local or provincial pressures. We may be poised for a large increase in availability in Canada. Alcoholic beverages have been introduced into grocery and corner stores in some provinces, and maybe in others. International treaties may demand the introduction of cheaper alcoholic beverages. Also, the monopoly system may be threatened by gradual or sudden moves toward privatization of retail stores. Lastly, because of the decline in alcohol sales, governments are receiving less revenue than previously. This has led some provinces to reduce the prices of alcoholic beverages or at least not keep them increasing with inflation. If government revenues drop further there may be pressures to further reduce prices and hence increase alcohol consumption.

Summary

Most alcoholic beverages consumed in Canada are produced here. Almost all beer and spirits are locally made, but much wine is imported. Taxes on alcoholic beverages are make up 53 per cent of the price of beer, 69 per cent of the price of domestic wine, and 82 per cent of the price of spirits. Just how these taxes are set at the federal and provincial levels is difficult to untangle. Alcohol controls other than price are established by both federal and provincial governments. Most of these controls are provincial, but the federal government controls importation, media advertising, and a variety of matters affecting purity and packaging. All provinces except Alberta operate a full or partial monopoly system for controlling the production and sale of alcoholic beverages. Provinces regulate the types of beverages, where they can be sold, and who can buy them during what hours. Minor changes are frequently made in the regulations on alcohol availability. Since the Second World War, alcohol availability has greatly increased in Canada as all provinces have moved from a puritan to a liberal approach. Privatization of alcohol distribution is coming slowly to Canada, although its impact is not yet clear.

Chapter 9

The Effects of Alcohol Control Policies

All countries have policies and practices for controlling access to alcoholic beverages. Canadian laws and regulations governing the availability and use of alcohol were reviewed in Chapter 8. The present chapter is concerned with what changes in controls the governments can employ and how these controls affect alcohol consumption and alcohol-related problems. Studies of the effects of price on consumption will also be reviewed. In this chapter we focus our attention on studies done in Canada, as they are the most likely to be helpful in future debate on alcohol policy.

The Density of Retail Outlets for Alcohol

While some Canadians make their own wine and beer, most alcohol consumed in Canada is purchased in stores, bars, and restaurants. Both the stores and the licences for the bars and restaurants are regulated by provincial governments. The numbers of all types of retail outlets have increased greatly since the Second World War. There have also been changes in the character of bars and taverns, such as improved decor and the introduction of games such as pool and darts and, in some, live entertainment. Many people assume that the increase in numbers of retail outlets has led to increased alcohol consumption. However, the evidence for a cause-and-effect relationship is mixed, and a clear conclusion is difficult to make from Canadian studies.

An interesting study of two Ontario rural municipalities conducted in 1966 showed that purchasing habits for home consumption were unrelated to retail availability (Popham, Schmidt and deLindt, 1978). The two municipalities were about the same size, but only one had a liquor store. People who lived in the municipality with no liquor store had to travel several miles to purchase alcohol. However, the purchasing habits of the residents of the two communities were very similar. In both communities, about 25 per cent of people bought alcohol at least once during the month of the study, and about five per cent made at least four purchases. Clearly the inconvenience of having to drive several miles to the nearest liquor store did not affect how often people bought alcohol. Whether the inconvenience of leaving town to purchase alcohol led to larger purchases on any one occasion is not known. However, as the proportions of more frequent purchasers (four times a month) were equal in the two municipalities, residents of both places were probably equally likely to purchase alcohol as needed or in bulk.

One study has examined the effects of introducing on-premise drinking in Owen Sound, Ontario (Smart and Docherty, 1976). Before the opening of three licensed lounges in 1973, Owen Sound had only beer stores selling for home consumption. This reflected a long history of "dry" sentiment in the town, located on Lake Huron. Traffic accidents and impaired-driving charges in Owen Sound for two years before and two years after the opening of the licensed lounges were examined. Similar statistics were also obtained for the town of Collingwood, the closest city on the lake, about 80 km from Owen Sound. Collingwood had both on- and off-premise alcohol outlets for many years and no significant changes in number or type of outlets during the study period. There were no obvious changes in the frequency of alcohol-involved traffic accidents and impaired-driving charges in Owen Sound after the advent of on-premise consumption.

The effects of increasing retail outlets was also considered in a Canada-wide study (Barnes and Bourgeois, 1977). The relationship between numbers of outlets and consumption was different for specific beverages. During the period of study (1951-1974) wine consumption was positively related with increases in the number of retail outlets. However, for spirits there was no relationship between consumption and the number of retail outlets, while for beer the relationship between retail outlets and consumption was actually negative. It is, of course, difficult to say whether changes in public taste accounted for the findings rather than changes in outlets.

A provincewide study in Ontario also examined numbers of on- and off-premise alcohol outlets in relation to alcohol consumption and alcohol-related problems in Ontario counties (Rush, Gliksman and Brook, 1983). The study found that rates of alcohol-related problems (illnesses, traffic accidents, and deaths) reflected differences in per capita consumption and that per capita consumption was itself related to the numbers of retail outlets. The authors concluded that governments could reduce alcohol-related problems by restricting the number of retail outlets.

Temporary closures of retail outlets during strikes have provided other opportunities to study the effects of outlet availability on consumption. Two studies found that public drunkenness decreased when retail stores were closed by strikes, but no other consistent effects of strikes on alcohol problems have been reported (Smart, 1977; Reid, 1976). The impact of strikes on per capita consumption was also considered in a Canada-wide study conducted by two researchers at McMaster University in Hamilton, Ontario (Johnson and Oksanen, 1974). The results suggested, not surprisingly, that sales of those beverages most affected by strikes do decrease. However, sales of other beverages increase, as do sales in other provinces of beverages affected by the strikes. Another Ontario study also showed that during two Ontario beer strikes Ontarians drank more wine and spirits (Single, 1979).

In three northern communities, there have been attempts to control drunkenness and assaults through restrictions on the retail availability of alcohol (Smart, 1979a). The three communities were Frobisher Bay (Iqaluit), a town of some 2,000 residents on Baffin Island; Rae-Edzo, two connected communities totalling 1,100 people on the North Shore of Great Slave Lake; and Fort Resolute (population 497) on the south shore of Great Slave Lake, about 950 km north of Edmonton. Each of these communities had a high frequency of alcohol-related arrests and assaults.

In Frobisher Bay, the sole package store selling alcohol was closed by the authorities in 1976, although four other outlets remained open (a hotel, a private club, a service club, and a liquor warehouse supplying outlying areas by mail order). Alcohol also continued to be available through mail order from Montreal or Yellowknife.

Rae-Edzo introduced total Prohibition in 1974 after many previous attempts to ration alcohol had failed to do much to reduce alcohol problems among local people. From that time, there were no legal retail outlets for alcohol in the community. All supplies had to be ordered or obtained from Yellowknife.

Fort Resolute also introduced total Prohibition in 1974, and in 1975 it also limited the amount of alcohol that any individual could legally possess at any one time. As in Rae-Edzo, the residents of Fort Resolute could continue to obtain alcohol by mail order or drive 96 km along a dirt road to the nearest liquor store.

Changes in the local retail availability of alcohol had no measurable effects on drunkenness, alcohol-related assaults, or impaired driving convictions in Rae-Edzo or in Fort Resolute. However, the new restrictions did have a considerable impact in Frobisher Bay. There, arrests for public drunkenness and alcohol-related assaults declined after the liquor store was closed. Frobisher Bay, unlike Rae-Edzo and

Fort Resolute, is not connected to any other towns by road. When the liquor store closed, residents of Frobisher Bay who wanted alcohol for home consumption had to arrange for supplies to be flown in from the south. Since this study, many northern Native communities have banned alcohol, but no study has been made of the effects of these measures on consumption. Most other Canadian studies have considered only small differences in numbers of retail outlets. Alcohol was more or less readily available in the various conditions studied, and differences in availability were not dramatic. As such, the effects on consumption could not be expected to be very large or easily detected.

Earlier studies of availability mostly dealt with the period up to the mid-1970s when consumption was mostly increasing. In Chapter 4 we showed that alcohol consumption has been declining in Ontario since about 1975 (Smart and Mann, 1995). However, this decline is not due to any reduction in the number of outlets. In fact, we have shown that the number of on-premise drinking establishments increased by 134 per cent in Ontario between 1975 and 1991, chiefly because far more restaurant licences were granted. The number of liquor stores also increased by 16 per cent, about the same rate as the population growth. However, alcohol consumption decreased by 19 per cent and daily drinking declined by 38 per cent over the same period. It is clear that people have been drinking less in Ontario despite large increases in the number of places to drink, raising serious doubts about how important the physical availability of alcohol is in influencing how much people drink.

Types of Outlets for Alcohol

Several Canadian studies have considered various characteristics of retail outlets in relation to alcohol consumption. For example, two researchers found that the opening of a new cocktail lounge in a small Saskatchewan community had no effect on alcohol consumption (Dewar and Sommer, 1964). The new lounge, which allowed both men and women to drink on the premises, replaced an older "men-only" bar.

Another study on types of outlets, conducted in Toronto (Smart, 1974), compared the self-reported purchasing habits of customers in two liquor stores. In one store, customers had to fill out order slips to get bottles from store clerks. The second store was self-serve. The people who shopped in the self-serve store reported more "impulse buying" and more drinking in the previous week than did those who used the clerk-service store. Customers of both stores also said that they were shopping at the store that was most convenient to their home or work.

Finally, two studies have looked at the effects of bar characteristics on drinking and drinking problems. One of these studies was done in Alberta and was concerned with the effects of dancing and games (Ratcliffe, Nutter et al., 1982). It was found that people who danced or played games drank more slowly than those who did not. However, they stayed longer and drank more.

In the second study, two observers spent two to three hours each observing at 185 Vancouver bars and made ratings of the bars' physical characteristics, atmosphere and the behavior of bar staff and customers (Graham, 1984). Special attention was paid to customers' apparent levels of intoxication and aggression. Bars with the fewest incidents of aggression and whose patrons seemed to get less intoxicated were smaller, did not have tables arranged in rows, had a "theme" decor, were clean, well maintained, and dimly lit and had no dancing or other activities. They also had recorded music, food, good ventilation, low noise levels, fewer patrons, female and friendly bar workers and high expectations of their clients. Intoxication and aggression were lowest in bars where patrons did not swear, fondle each other, move about, or be familiar with strangers.

This study's results may reflect the tendency of people to select bars where they can behave as they wish, but bar characteristics may have a direct effect on the way customers behave. Certainly this has been the assumption of legislators who, over the years, have laid down diverse regulations for licensed establishments. The possibility that bars could be deliberately designed to minimize undesirable behavior has also been considered. In the 1970s, the Alberta government introduced regulations with precisely this possibility in mind. The new regulations required large beer halls to be partitioned and bar sizes to be reduced. Improvements in decor and the introduction of dress codes were also required. Unfortunately, no rigorous evaluations of the effects of these new regulations have been made. However, a news magazine account of the effects of the changes indicated that bar managers and other officials were pleased with how the changes affected customer behavior (Zwarun, 1978).

In the past few years we have seen large numbers of new alcohol outlets develop, but there is little or no research on what effect they may have had. One of the most striking changes is in the number of patio restaurants with outdoor seating. Also, theatres and some museums can serve alcohol to patrons. People are allowed to drink on their private boats, and large pleasure or "party" boats may carry hundreds of people who are drinking. There are also wine bars and brew pubs serving their own beverages. Many people increased their visits to the United States from border cities greatly in the 1980s and many of them brought home alcoholic beverages at low prices. More recently, brew-your-own beer and make-your-own-wine shops have proliferated, at least in Ontario. All of these developments have made alcohol more accessible and some have made it much cheaper. Unfortunately, we know very little about how this has affected people's drinking in Canada.

Recently, many municipalities have allowed beer to be sold in the stands during hockey games and other sports events, despite fears that this might lead to more violence. A study done in Toronto (West, Dick et al., 1994) found that after beer was sold during hockey games the pattern of violence changed, but the total number of incidents did not. Probably some drinking occurred illegally before the new licensing, which may have made any effect of the change difficult to detect.

Another type of alcohol marketing that has been studied is "happy hours." These were first allowed in most provinces in the early 1980s. They involve two-for-one offers or price reductions at certain times of the day. Because price is usually an important influence on consumption, it was expected that happy hours would make people drink more. However, when they were banned in Ontario in 1984 there was no effect on alcohol sales data or charges for impaired driving (Smart and Adlaf, 1986).

The most important recent study about types of outlets comes from the national survey in 1988. Single and Wortley (1993) have shown that Canadians who drank in bars and taverns and at parties had higher levels of drinking and more drinking problems. This does not mean, however, that such drinking venues cause problems. It is likely that heavy drinkers prefer to do their drinking in those situations. Lighter drinkers prefer to drink at home, with their family or friends.

Hours of Sale

Despite the long-standing belief that opening hours, especially of outlets for on-premise consumption, affect public drunkenness and impaired driving, there has only been one relevant Canadian study, done in 1962 in Toronto (Popham, 1962). The researcher plotted hourly patterns of drunkenness arrests between 8 a.m. on a Monday and 8 a.m. the following Sunday. The results suggested that there were more arrests during tavern opening hours than when the taverns were closed. However, a similar pattern of arrests was found between 8 a.m. on Sunday and 8 a.m. on Monday — a period when all the taverns were closed.

Alcohol Content and Type of Beverage

The type of beverage consumed does not seem to influence physical problems such as liver cirrhosis. What matters is the amount. For most purposes, one 12-oz. beer (five per cent alcohol by volume) can be considered equivalent to a five-ounce glass of wine (12 per cent alcohol) or one 1.5-oz. shot of liquor (40 per cent alcohol). Each of these drinks contains about 13 mL of pure alcohol and, all else being equal, drinkers of the same number of standard drinks of any type of beverage over a given period of time will be equally impaired. However, impairment will come faster with the higher-alcohol beverages unless they are diluted to the strength of beer.

Despite this, many people believe that beer and wine are relatively benign compared with "hard liquor," probably because the same volumes of different beverages do produce different degrees of intoxication. (A litre of whisky contains eight times as much alcohol as a litre of beer.) Thus, it is necessary to drink a greater volume of beer or wine to reach the level of intoxication obtained from spirits.

Despite the findings for cirrhosis, beverage type is related to some other alcohol problems. For example, a study of college students in the United States found that the beer and spirits quantity-frequency indices were correlated with "serious and uncommon" problems such as hallucinations, blackouts and morning drinking. The wine quantity-frequency index was not correlated with problems (Kilty, 1990). A 1975 study found that beer consumption was related to self-reported drinking-driving accidents, but wine consumption was not (Perrine, 1975). Similarly, Berger and Snortum (1985) in a survey of licensed drivers of all ages in the United States found that those who typically drank beer drank to higher levels of intoxication and were more likely to drive after drinking. Beck and Summons (1987) surveyed 2,000 high school students in Maryland and found that beer and liquor drinking were more common among drinking-drivers.

In our study of Ontario students in 1991 we found that few students drink beer, wine or spirits exclusively and that most drank more than one beverage type (Smart and Walsh, 1995). Those who drank wine only and beer and wine only were less likely to drink five or more drinks on a single occasion. Wine drinkers reported the lowest levels of heavy drinking. Beer and liquor drinkers and drinkers of all three beverage types had the highest rates of drinking, heavy drinking, drinking and drug problems, drug use and delinquency. However, when age, gender and quantity/frequency were taken into account, most beverage-type differences disappeared. Still, beer and liquor drinkers and drinkers of all three beverages did more often drink five or more drinks on an occasion and had higher rates of delinquency.

The beverage of moderation for adolescents appears to be wine, if taken exclusively. In our bivariate analyses of the Ontario student sample, wine drinkers had the lowest rates of heavy drinking, drinking problems, drug use and delinquency. However, this probably reflects the fact that this group drinks less overall. Also, wine drinking may occur in family situations such as meals, festivities and parties which are closely monitored. Some wine drinkers probably belong to ethnic groups that traditionally drink wine at religious events. Exclusive wine drinkers were, however, predominantly female and they were the smallest category of drinkers (one per cent).

Two Canadian studies have also examined deliberate attempts to regulate the availability of specific beverages. The first examined the effects of removing 60 low-priced Canadian fortified wines from stores in northern Ontario (Giesbrecht and Macdonald, 1981). The intent was to reduce alcohol problems among native people, who were among the major consumers of domestic fortified wines. At the time they were withdrawn, these wines were the least expensive (by volume of absolute alcohol) of all alcoholic beverages. Sales of various types of drinks were monitored in the 10 communities where the fortified brands were removed and in 18 communities where they continued to be available. There was a slight drop in overall alcohol sales in the delisted communities and a slight increase in the purchase of delisted beverages from neighboring communities. However, the most striking results were dramatic increases in the sale of domestic red and rose wines and in sales of vodka and Liquor Control Board alcohol. Both can be used to fortify regular wines, and there was evidence from community workers that this is exactly how they were used.

Another possible approach to regulation would be to lower the alcohol content in the beverage. A study of the users of light beer provides some insight into how such an approach might affect overall consumption (Whitehead and Szandorowska, 1977). This study, conducted in London, Ontario, in the summer of 1975, focused on Labatt's Cool Spring — a low-alcohol beer introduced in the spring of 1975. Drinkers of Cool Spring were asked about their drinking habits in a range of situations before and after their first use of Cool Spring. While many of those interviewed reported drinking Cool Spring when they would previously have had regular beer or other alcoholic drinks, about seven per cent said that they drank Cool Spring when they would previously have abstained. Cool Spring was most likely to be drunk as an alternative to abstinence at mealtimes.

The introduction of light beers thus seems to have increased the range of situations where alcohol is used. If the range of situations in which alcohol is used increases, the frequency of some alcohol-related problems is also likely to increase. Unfortunately, neither this study nor any other tells us what volumes of Cool Spring or other alcoholic beverages were used by drinkers who either added light beer to their diets or used it as a substitute for other drinks. It is possible that consumption per drinker of pure alcohol decreases when drinkers switch to low-alcohol beer, even if some also then drink on more occasions. More research is clearly needed, since many users of low-alcohol beer are attracted to it because of their concern with health and diet, and these beers are a growing part of the market.

Drinking Age

All industrial societies have established lower age limits for the legal purchase and possession of alcohol. In recent times these limits have ranged from 14 to 21 years. Until the early 1970s the legal age for the possession of alcohol in Canada was 21, except for Quebec, where it was 20. However, by 1972 all 10 provinces and the northern territories had reduced the legal drinking age to 18 or 19.

Initially, there were many arguments in favor of lowering the drinking age in Canada along with the age of legal majority. One concerned the supposedly "inoculating" effects of an early introduction to alcohol. In brief, this argument holds that if young people are allowed to drink, they will not see alcohol as a forbidden fruit, but rather learn to use it responsibly and thus avoid alcohol problems. This contrasts with the view of young people as especially vulnerable to the damaging effects of alcohol and likely, through irresponsibility or ignorance, to develop inappropriate drinking habits that will be difficult to modify in later years. Proponents of these opposing views naturally made different predictions about the effects of lowering the drinking age. Others argued that the effects would be minimal because the new law would mainly allow large numbers of 18- to 21-year-olds who were already drinkers to drink legally.

Several studies examined the effects of the legal drinking age reductions in Canada and the United States (Smart and Goodstadt, 1977; Vingilis and Smart, 1981). The results show that lowering the drinking age increased both drinking and alcohol-related traffic accidents among young people. The largest increase in young people's drinking was in on-premise consumption rather than in off-premises sales or drinking with families. In Canada, the overall effects of the change in the legal drinking age varied from province to province, with Ontario and Saskatchewan appearing to experience more problems than elsewhere. Canadian studies have not shown that lowering the drinking age increased educational, family, or legal problems among young people or led to increases in rates of admission to alcoholism treatment programs (Vingilis and Smart, 1981).

One reason why age laws are relatively ineffective has to do with the easy access young people have to alcohol. After the age law was increased to 19 in Ontario, special age-of-majority cards were required for young people to buy alcohol. One study showed that nine per cent had a legal card, but 74 per cent drank alcoholic beverages. About five per cent of students borrowed cards from other people. Those who had cards were more likely than others to be heavy drinkers and to have drinking problems. Of course, most student drinking occurs without needing to buy alcohol. Most get alcohol from parents' supplies, from friends' purchases or by attending parties.

When Ontario raised the drinking age from 18 to 19, there were few short-term effects on alcohol consumption among high school students. In one Ontario town, no effects were found on the incidence of young people charged with under-age drinking or on alcohol-related driving convictions and fatalities involving young people. However, in the United States there are many studies that show that raising the drinking age to 21 lowered rates of heavy drinking and drinking driving among young people.

Drinking ages no longer seem to be an important issue in Canada, either for politicians or the general public. There is still some question about when young people should be permitted to drink under parental guidance. For example, it is uncertain whether an early introduction to alcohol necessarily reduces the chances of later alcohol-related problems. In a review of Canadian studies relevant to this and other topics concerning young people's drinking, some interesting results emerged (Whitehead, 1983). In general, young people who drink at home are more likely than others to also drink away from home and to drink larger quantities away from home. This result is inconsistent with the "inoculation" hypothesis, as are findings of several studies showing that people with alcohol problems start drinking at an earlier age than drinkers without such problems. First use of alcohol away from home predicts later problem drinking for some, whereas many other problem drinkers report first drinking with their parents. The reviewer concludes that the bulk of the evidence indicates that "social groups that initiate young people into the use of alcoholic beverages at a young age in a family setting and where alcoholic beverages are a normal part of everyday life tend to have a high average level of consumption and high rates of alcohol-related damage. France and Italy may be the best known of such societies, but they are not the only ones."

Advertising and Promotion

The frequency with which alcohol advertisements appear on television and radio and in print media has caused concern that these contribute directly or indirectly to alcohol consumption. Some worries have been the possibility that alcohol advertisements foster tolerance toward alcohol use in a wide range of situations, contain "subliminal" messages that influence the unconscious, and portray drinkers as attractive, healthy, and mature. Some people say that the ads for some beverages are aimed especially at young people, who are the most vulnerable to advertising's effects. The possibility has also been raised that alcohol advertisers may also affect how the media deal with alcohol-related issues in news stories or other programs.

Concerns with the "power" of alcohol advertising find expression in the host of regulations that each province has enacted respecting the content, size, and location of alcohol advertisements (see Chapter 8). Alcohol manufacturers and distributors find these regulations frustrating, especially because they vary from province to province. In recent years there has been much concern about "lifestyle" advertisements that show drinkers doing exciting things. These ads are supposed to be especially seductive. In Ontario, guidelines for such advertisements were drawn up by the provincial government in the late 1970s, but there is no evidence that they have had an effect on alcohol consumption.

The actual effects of alcohol advertising on alcohol consumption and on alcohol-related problems are very difficult to determine. Special problems are posed by the ubiquity of alcohol advertisements, the difficulty of getting hold of advertising and sales data from manufacturers, and the way levels of advertising increase as sales increase. This last problem makes it hard to show any causal relationship. Some researchers feel that these problems are so great that it is impossible to determine if alcohol advertisements influence alcohol sales and consumption. The advertisers maintain that advertising does not increase consumption, but rather influences brand preference.

Two studies by the present authors concerned the effects of new alcohol-advertising regulations in British Columbia and Manitoba (Smart and Cutler, 1976; Ogborne and Smart, 1980). The B.C. study focused on the effects of a temporary ban on alcohol advertisements in local newspapers and magazines and on local radio and television. The ban came into effect in 1971 and ran for 14 months. It was never particularly supported by B.C. residents and, not surprisingly, was opposed by the alcohol producers and the advertising agencies.

The Manitoba study (Ogborne and Smart, 1980) looked at the effects of the removal in 1974 of all beer advertising from the electronic and print media produced within the province. Wineries and distillers continued to advertise in the print media, and some television commercials for wine were also allowed. As was the case during the B.C. ban, broadcast and print material originating outside Manitoba could not, of course, be controlled by local regulations.

Neither study showed that the changes in advertising policies had any effect on per capita consumption or on the consumption of particular beverages. If anything, beer sales in Manitoba increased after the advertising ban, but there was a similar increase in beer sales in Alberta over the same period.

The most recent study of an advertising ban was done in Saskatchewan. In 1983 a 58-year alcohol-advertising ban was rescinded. Makowsky and Whitehead (1991) examined monthly sales of beer, wine and spirits. They found that sales of beer increased and sales of spirits decreased, but there was no overall effect on total consumption of wine. They concluded that alcohol advertising does not influence the overall level of consumption.

Several experimental studies have been done in Canada with alcohol ads including one conducted in a Toronto shopping mall (Kohn, Smart and Ogborne, 1984). Male shoppers between the ages of 19 and 45 were asked to evaluate samples of one of two types of advertisements for an imported beer or to answer general questions about the mall. One type of advertisement showed young people enjoying themselves while drinking the beer, and the other stressed the beer's qualities and showed no people. These two advertising styles are sometimes termed "lifestyle" and "tombstone" respectively.

Each person interviewed was given a $5 voucher valid for immediate use at one of the mall's licensed restaurants. At the insistence of the research ethics committee, the vouchers could not be used for the purchase of alcoholic beverages, but only for the food portion of the bill. However, the management of the restaurants agreed to staple bar bills to the vouchers of any customers who purchased alcohol with their meals. Six to twelve weeks later, 84 of the original 163 people who used the voucher were contacted by telephone for an ostensibly independent survey about drinking. Neither the lifestyle nor the tombstone advertisements had any impact, immediate or delayed, on alcohol consumption in total or on use of the advertised beer. Those interviewed did not differ in their evaluations of the two styles of advertisements used.

Perhaps the limited exposure to these advertisements could not have much effect on consumption, given the mass of alcohol advertisements to which people are normally exposed. However, the advertisements used were for a beer that is rarely advertised in Canada and the exposure to these advertisements during the study was unusually intensive. Each of five ads had to be rated and then all five had to be ranked in order of preference. The fact that the vouchers could not be used to purchase alcohol was acknowledged as an unfortunate limitation.

Two other studies examined the effects of television advertising of alcohol on consumption in party situations. In the first, male university students were exposed to beer advertising while able to drink (Kohn and Smart, 1984). Subjects, who volunteered for a social experiment, watched indoor soccer matches and filled out a questionnaire. They watched televised matches that contained no beer commercials, four commercials or nine commercials at the rate of about three per half hour. Beer, non-alcoholic beverages and snack foods were available during the sessions, each of which involved four to six students. The experiment was held in the evening in a faculty lounge often used for parties. It had a bar, comfortable over-stuffed chairs and subdued lighting. Actual consumption by individuals was monitored. Over the course of the experiment the amount of advertising had no effect on beer consumption, although there were suggestions of short-term effects. Delayed access to beer led to compensatory consumption when it became available, but had no effect over the whole experiment. The study attempted to created a natural party situation and both the drinking and television advertisements were similar to those students normally experience. However, there was no check on whether the subjects accepted the explanation for the study given by the experimenter.

Suspiciousness on the part of subjects was important in the next study of alcohol advertising done by Kohn and Smart (1987). It involved women college students who viewed videotapes, with or without wine commercials, in the same party situation as for the male "beer study." Women college students volunteered for a project represented as an evaluation of televiewing tastes. The experiment took place in the evening in the same faculty lounge as used for the "beer and males" study. Subjects in groups of four to six viewed a soap opera, a popular music program and a beauty contest containing zero, three or nine commercials for wine. They filled out questionnaires after the first two. Wine and snack foods were available.

Unlike the male study, this one involved a debriefing session and a questionnaire assessment of suspiciousness about the true purpose. Twelve of 66 subjects were suspicious that the study was not as presented (most thought it was about reactions to commercials generally or about group interaction and conformity). Women shown nine commercials drank more wine than those shown three. However, there was no difference between the zero commercial condition and the three and nine condition. These results, however, disappeared when the "suspicious" women were dropped from the analysis. Probably, personality factors make suspicious women resist mild persuasive tactics, but accede to strong ones. Given these results it is unfortunate that there were no "suspiciousness" estimates for the male beer study. There is a need for more studies that take into account a variety of personality variables such as persuadability and suspiciousness.

Only one Canadian has examined the effects of alcohol content in both programming and advertising (Sobell, Sobell et al., 1986). Male college students were asked to volunteer for experiments with prime time television programs and taste tests of light beer. They viewed video tapes of a *Dallas* segment doctored to include: (i) four alcohol advertisements and 14 drinking scenes in the program content or no drinking scenes; (ii) no drinking scenes and alcohol advertisements only; (iii) alcohol scenes and no alcohol advertisements; and (iv) a control condition for the various combinations of no alcohol scenes and non-alcoholic beverage or food advertisements. Subjects performed a taste test for light beer and the amount consumed was measured. Neither the drinking scenes nor alcohol advertisements had any effect on the amount of beer consumed. However, there was an interaction between program type and advertisement type that the authors construed as unimportant.

As with most experimental advertising studies, the Sobell, Sobell et al. study (1986) was not ideal. There was no party situation and subjects were tested individually. The tests were held in late morning or early afternoon — times when many students would not be in the mood for drinking — and, as it was conducted in the Addiction Research Foundation, many subjects may have been distrustful of the rationale given.

In summary, a variety of experimental studies have examined the short-term effects of alcohol advertisements, but none is ideal methodologically. The experimental drinking settings were sometimes artificial and contrived efforts to parallel human drinking situations. However, the consistently negative results, along with the results of economic and advertising ban studies, do suggest that concerns about alcohol advertising may be exaggerated. Alcohol advertising is probably one of the weaker variables with which to control alcohol consumption (Smart, 1988).

This is not to say that alcohol advertising has no effects, but only that such effects have not been convincingly demonstrated. Perhaps our methods are too crude and the effects are more subtle than at present imagined. Thus it may be that some types of advertisements do affect some types of people in some types of situations. For example, children may over the course of their growing years be influenced by lifestyle advertisements. Some people may learn from alcohol advertisements that they would be more confident, aggressive or talkative if they drank. At present, however, these are speculations that need to be researched.

The Price of Alcoholic Drinks

The price that Canadian consumers pay for alcoholic drinks is determined by many factors other than manufacturing costs and profits. Both wholesalers and retailers add their markups. On-premise drinks may be double or even quadruple the off-premise price. However, a major influence on the real price of alcoholic drinks, at least at package stores, is the imposition of various taxes.

Taxes on alcoholic drinks are almost universal and have been justified both as a means for generating government revenues and as a way to regulate alcohol consumption. This latter justification is not often voiced in Canada in recent years, as it sounds antilibertarian and paternalistic to modern ears. Nonetheless, the prices that consumers pay for alcohol often do have a significant impact on consumption and, in consequence, on some alcohol-related problems. Governments may therefore be able to use their powers to regulate the real price of alcohol through taxation to achieve public health goals.

In recent times the first researcher to be much concerned about the relationship between the price of drinks, consumption, and alcohol problems was J. R. Seeley (1960). While working at the Addiction Research Foundation in Toronto, Seeley showed that in Ontario from 1921 to 1956 there were remarkably close relationships between the price of alcohol as a percentage of the average consumer income, per capita alcohol consumption, and liver cirrhosis death rates. In fact, per capita alcohol consumption and liver cirrhosis death rates have followed very similar courses in Canada up to about 1982. Both generally showed a strong inverse relationship with the real price of alcohol. This is confirmed in a later study for the years 1955 to 1982 in both Ontario and Michigan (Rush, Gliksman and Brock, 1983). Studies conducted in other countries have produced similar results, confirming that price plays an important role in influencing both consumer demand for alcohol and the overall rate of consumption.

Although the price of alcohol and the rates of alcohol problems have been traditionally related, this may not always hold true. During the decline in alcohol consumption in Ontario in the 1980s, studies at the ARF also registered even larger declines in liver cirrhosis and drinking and driving rates. Most alcohol-policy experts would have expected that alcohol prices in Ontario went up. However, over the period 1975-1991, real prices were essentially stable and thus price did not contribute to the declines in problems.

The effects of price changes on the broad range of people who drink alcoholic beverages are complex. Price cannot be considered in isolation from the many social and economic factors that influence consumption. For example, wealthy consumers will be less affected by increases in the price of alcohol than will those living on lower incomes. Also, there are ways to avoid the effects of high alcohol prices. Those who buy alcohol in airport duty-free shops or in U.S. border cities will pay lower prices than the average Canadian. Heavily dependent drinkers may give up other pleasures if the price of alcohol increases and hence their total recreational spending may remain the same. Also, in situations where alcohol is not readily available, the effects of price or income changes may be less than in situations of greater availability. Finally, the home production of beer and wine or use of brew-your-own stores allows people to reduce the cost of alcohol. The opening of new stores may be stimulated by increases in the price of commercially produced alcohol. Drinkers might increase their consumption if inexpensive home-made product can be made easily and with little capital outlay.

The relative price of alcohol is only one of many factors influencing overall consumption. Also, the precise relationship between relative price and consumption cannot be specified. Of course, price is one of the few factors that could be manipulated by governments concerned with public health aspects of alcohol consumption. Most other determinants of alcohol consumption that could be influenced by governments hold little promise as public health measures.

Summary

Studies of small changes in the numbers of retail stores show no consistent effects on consumption or alcohol-related problems. However, significant changes in retail availability, especially of outlets selling alcohol for home consumption, can influence consumption and alcohol-related problems. Some characteristics of bars are associated with higher levels of intoxication and aggression among bar patrons. The effects of retail stores' hours of operation on consumption and alcohol problems have not been studied extensively in Canada. Attempts to reduce consumption and alcohol problems by regulating the physical availability of specific beverages have not always been successful. In the early 1970s, lowering the drinking age resulted in increased alcohol consumption and increased traffic accidents among young people. Canadian studies have not shown that alcohol advertising has a direct effect on alcohol consumption. However, per capita alcohol consumption has increased in Canada with decreases in the relative price of alcohol. Reductions in alcohol consumption can, however, occur with no increase in price and a large increase in availability of alcohol as found recently in Ontario.

Chapter 10

Priorities For The Prevention of Alcohol Problems

Some First Thoughts

Prevention has usually been a poor stepsister to treatment when it comes to alcohol problems. Virtually all cities and towns in Canada have some treatment for alcohol problems, including options such as specialized treatment programs, Alcoholics Anonymous or general hospitals. However, few have any major, sustained prevention programs in the community. In some provinces, such as Ontario and Alberta, alcohol education was mandated for schools in the 1990s so progress is being made.

Of course, Canadian society needs to prevent alcohol problems. We will not solve the difficulties merely by providing treatment for alcoholics, often called tertiary prevention. What is needed is primary prevention — prevention of new problem drinkers and new problems — and secondary prevention, or the early detection of problem drinkers and interventions to minimize the chances that they will get worse. It is expensive and probably ineffective to attempt to control serious socio-medical problems brought on by alcohol merely by treating affected cases.

A first need in prevention is to define what it is that we wish to prevent. In general, we wish to prevent problems of the sort described in Chapter 7. The main concerns are death and injury from traffic and other accidents and long-term medical consequences of alcoholism, such as liver cirrhosis, neurological disease, and premature death.

We should emphasize that we do not wish to prevent the drinking of alcohol in all its forms. There are some good reasons for careful moderate drinking — especially mild psychological and social betterment. The possibility of some cardiovascular protective benefit from moderate use still exists. In addition, the alcohol beverage industry provides jobs, incomes for farmers, and many useful taxes for governments. Most social drinking leads to no important social problems and clearly has some benefits for most people. An equally important reason for not trying to prevent all drinking is the impracticality of the attempt. Few people in Canada seriously believe that Prohibition can return, many would doubt that it could ever be effective, and others would be against it in principle. Virtually none would support it, and without public support such policies can never be effective.

It is doubtful whether it is worth trying to prevent even harmless public intoxication. Prevention of drinking that is causing no harm or damage becomes more a question of moral belief than of public health. Of course, we would seek to prevent any "drinking with damage" — social, psychological, or biological harm to the drinker and to others.

We argue here for an emphasis on primary prevention (prevention of the original problem) as well as secondary prevention (early detection and treatment of problems). Both are much preferred to tertiary prevention or treatment, which tends to be both expensive and not very effective. If we accept the need for primary and secondary prevention, then changes in public policy and strategies about alcohol are required.

Public Policy and Prevention of Alcohol Problems

Models of Prevention

No proven method for preventing most alcohol-related problems exists except for solutions so drastic that few countries would accept them. The largest reductions in drinking and drinking problems ever found have occurred in wartime or during Prohibition (Bruun, Edwards et al., 1975). Of course, neither of these represent alcohol control options for modern governments. Consequently, governments are left with a variety of questionable experiments and untried models for prevention.

The most widely accepted models are the "distribution" model and the "socio-cultural" or "integration" model (Popham, Schmidt and deLint, 1978). The notion that the problems of alcoholics have little to do with drinking, but are the result of psychiatric or physical predisposing factors in the alcoholic, is not taken very seriously any longer. It demands a division of drinkers into social drinkers and alcoholics. Since the claim is made that rates of drinking have nothing to do with alcoholism, this view has been popular with the manufacturers of alcoholic beverages. A serious problem is, of course, that any such clear-cut division has never been found in Canada and rarely elsewhere. Also, the major problems with most alcoholics appear to be unrelated to underlying psychiatric or physical factors, but arise from their drinking too much over a long period of time. We have seen a movement away from a concern with "alcoholics" and towards a concern with alcohol-related problems such as drunkenness, drinking and driving, domestic violence, job loss and the like. Most of these problems occur at various levels of heavy drinking, not necessarily alcoholic drinking and, hence, we see less difference between social drinkers and those with problems.

The Distribution Model of Prevention

The distribution, or "single distribution," model began as a challenge to this notion of a clear division between social and problem drinking. It was originally developed by a Frenchman, Sully Ledermann, but after his death it was expanded and refined by scientists at the Addiction Research Foundation in Toronto. It has often been referred to as the "Toronto model," the "Canadian approach" or the neo-prohibitionist approach to prevention of alcohol problems. Much debate and research has followed the model, but we will not go into detail here. Some aspects of the model have now been accepted in a variety of countries and in official documents of the World Health Organization (Bruun, Edwards et al., 1975). Although the model still has its critics, they have diminished in both number and persuasiveness over time.

The main tenets of the distribution model for which there is most general support are:

- The general level of alcohol consumption in a population is closely related to the number of heavy users or alcoholics; the higher the per capita alcohol consumption, the larger the number of heavy users.

- A decrease in per capita consumption — especially a large one — will be accompanied by a reduction in the number of heavy users and those with problems.

- The overall availability of alcohol has an independent effect on the level of alcohol consumption; when alcohol becomes more available because of price or physical circumstances, per capita consumption will go up.

- Governments should stabilize or reduce per capita alcohol consumption by establishing control policies to increase prices and reduce availability. These methods will eventually reduce the proportions of heavy users and people with problems in the population (Bruun, Edwards et al., 1975).

Critics of the model deny that the relationship between consumption and heavy use is close enough to bother with, and that availability and price are not strong predictors of consumption (Parker and Harman, 1978). They also claim that the only problem of interest to "single distributionists" is liver cirrhosis, and that cirrhosis does not relate closely enough to other alcohol problems to be the only one considered. This last criticism has some value. Liver cirrhosis rates are not always closely associated with other problem indicators such as accidents, arrests for drunkenness, and hospital admissions.

Certainly one main problem with the single distribution theory is that it has not been given a serious trial. No government in Canada or elsewhere has taken its tenets seriously enough to create a social experiment. The requirement of the theory that per capita alcohol consumption be lowered for some years in order to see whether there is indeed an impact on heavy use and problems has never been tested in practise.

Governments in Canada make fleeting efforts to control consumption by increasing prices slowly or controlling advertising, but these are usually offset by increases in availability such as the licensing of more bars and restaurants, the lowering of drinking ages, or the sale of wine and beer in grocery stores. Without a serious experiment in social controls, we may never know for sure whether the distribution model is correct.

Another serious weakness of the theory may be that it does not predict well for situations where alcohol consumption is falling. Our research (Smart and Mann, 1995) showed that consumption in Ontario declined during the years 1975-1993. Alcohol problems declined even more steeply. However, this occurred at a time when prices for alcohol were stable and the number of outlets for alcohol greatly increased. Probably decreased consumption occurred because of changes in attitudes and lifestyles, but also as a result of increases in alcohol education and health promotion programs. There is clearly a need to rethink the theory that prices and availability control alcohol consumption.

The Socio-Cultural Model of Prevention

Another set of approaches to prevention has been defined as "socio-cultural," because they emphasize the social context of drinking and the backgrounds of drinkers. A major tenet of socio-cultural approaches is that the eradication of drinking problems requires changes in social norms about drinking. Adherents of the model believe that problems result from a lack of norms about what is "safe" drinking and a lack of social pressure not to exceed the safe level. Reference is often made to safe drinking among Jews, the French and Italians. Jews have both a low level of consumption (although most are drinkers) and few problems. However, Italians and the French, at least those living in Europe, have high rates of liver cirrhosis, but low rates of public intoxication.

Other problems exist with the socio-cultural approach. Some proponents have suggested that we should integrate drinking into everyday life — having wine with meals, for example — and make a clear distinction between drinking and drunkenness. However, if most people drank wine with most meals, there would be far more drinking overall. Also, studies of attitudes in Ontario have shown that most people already make a distinction between drinking and drunkenness and have negative attitudes toward drunkenness. Nevertheless, we still have many problems with alcohol. Unfortunately, we do not know what the safe level of drinking is for all types of problems. Efforts have often been made to establish maximum levels of drinking. Scientists at the Addiction Research Foundation (Ashley, Ferrence et al., 1994) now recommend that people drink no more than two drinks per day, that non-drinking days should be planned and that intoxication should be avoided. Pregnant women should not drink nor should people who are taking other medicines, those who operate vehicles or machinery, or those who have problems controlling their drinking.

Some people have suggested integrating the socio-cultural and distribution models by adopting policies that would limit per capita consumption while integrating alcohol into everyday life (Whitehead, 1975). The main suggestions seem to be for limits on alcohol advertising, which typically does not promote norms of moderate use, and a public education campaign against drunkenness and impaired driving.

Current Efforts to Prevent Alcohol Problems in Canada

Most of the efforts in Canada to limit alcohol problems have involved some nod toward the distribution or the socio-cultural approach. Most have been piecemeal efforts and of short duration, usually concerned more with changing attitudes than behavior. Often the focus has been on a particular group in society, such as young people or drinking drivers, rather than the population as a whole. Among the most promising approaches have been increased legal drinking ages, certain school alcohol programs, some drinking-driving programs and some selected controls on availability.

As discussed in Chapter 5, one study has shown that there are two groups whose attitudes particularly need changing — older, infrequent-drinking males and young males who drink heavily (Smart and

Liban, 1981). Older, retired, male infrequent drinkers often felt that those with drinking problems should take care of themselves, that the community had no responsibility and that treatment was ineffective. Younger male heavy drinkers frequently approved of heavy drinking in dangerous situations such as when driving, at parties or bars, and at sports or recreational events. Apart from these two groups, most people seem to require little attitude change, but young males clearly have the most dangerous attitudes.

Raising the Drinking Age

Although it was an important issue of the 1970s the drinking age issue seems to be largely dead in Canada now. In Canada, drinking ages in most provinces were reduced from 21 to 18 or 19. The same changes were made in about half of the states in the United States. The overwhelming evidence was that youthful drinking increased after those reductions and that problems such as drinking-related accidents also increased (Smart and Goodstadt, 1977). After the disillusionment with these reductions several provinces and states increased drinking ages again, but usually to 19, not 21, at least in Canada. In the United States, the federal government passed legislation that meant the states could not get highway repair funds unless they increased the drinking age to 21. All states have done that, while in Canada drinking ages are still 18 or 19. If we compare drinking levels among students aged 18 and 19 in Canada (Eliany, Giesbrecht et al., 1992) and the United States (Johnson, O'Malley and Bachman, 1992) we see that they are about the same. However, it is difficult to get comparable measures of heavy drinking or alcohol-related accidents on a national level.

There is reason to believe that an increase in the drinking age to 21 would reduce alcohol-related accidents in Canada. Many U.S. research studies have shown that the increase in drinking ages from 18 to 21 reduced drinking and driving fatalities (Stewart and Klitzner, 1990). Probably such increases would reduce levels of drinking and drinking-driving for people aged 19 to 21, but we don't have any relevant studies in Canada. In Canada such problems have declined rapidly among young people. For example, drinking and driving rates fell by almost 50 per cent among Ontario students between 1977 and 1991, although there has been a slight increase since then (Adlaf, Ivis et al., 1995).

It would be difficult to get public support or the political will to change the drinking age to 21. A recent survey in Ontario indicated that about half the adult population want the drinking age to be increased and half want to leave it the same. Typically, young people want the drinking age lowered; older people don't. The drinking age issue seems to have slid off the public agenda, and it would be difficult to make it an important issue again.

Controls on Availability

The evidence reviewed in Chapter 9 about availability and alcohol consumption suggests that few strong conclusions can be made about how alcohol controls should be established. Studies of controls on the numbers of sales outlets, types of outlets and type of beverages in Canada suggest little that can be used in a preventive program. Unless the changes in numbers of outlets or hours are very great, there would be very little impact on sales and hence on problems. Alcohol advertising has not been shown to affect sales (Smart, 1988) and thus is not a strong candidate for stricter controls. In any case, it is doubtful that major reductions could be made in numbers of outlets, hours, or volume of advertising.

The traditional choice for governments that want to control consumption is price — the real price of alcoholic beverages (Bruun, Edwards et al., 1975). As all governments in Canada strictly control the prices of alcohol through their monopoly systems, the concept is not new. We know that the largest portion of the selling price of alcohol in Canada is tax. However, the prices of alcoholic beverages have been set with government revenue, not alcohol problems, in mind. It could be argued that they should be set with one eye to public health, leaving the other eye to the needs of the treasury.

We also know that in real dollars, undiminished by inflation, alcoholic beverages are cheaper now than they were thirty years ago. It takes much less time for the average worker to earn enough money to buy a drink that it did in the 1950s. There is a strong case for governments in Canada to increase the price of alcoholic beverages considerably; however, this move would be most unpopular and is unlikely to be generally adopted. A more acceptable proposal is that governments be careful to maintain the current costs of alcoholic beverages in real dollars: that is, when inflation goes up, the price of alcoholic beverages should rise by the same amount, based on the amount of alcohol in them. In some years this would lead to small increases in prices, but in others the increases could be very large. For such a measure to be accepted by Canadian drinkers (and taxpayers), a large-scale education program would be required explaining the reasons for the increases in public health terms and the probable benefits to society. At least one survey showed that most people were willing to pay more for alcoholic beverages if it could be shown that higher cost would reduce alcohol problems (Goodstadt, Smart and Gillies, 1978). Of course, the education program would have to be persuasive and long-term.

A major problem with using price as the main alcohol-control variable is that in the last 15 years or so real prices of alcohol have not changed in Ontario. However, alcohol consumption and problems both decreased greatly. Clearly price is not the only factor controlling consumption.

Warning Labels on Alcoholic Beverages

There has been considerable debate in Canada about the need for warning labels on bottles containing alcoholic drinks. Several countries, such as Mexico, Colombia and the United States have warnings on the labels of alcoholic beverages. In Mexico and Colombia the warnings remind drinkers that alcohol (abuse of alcohol in Mexico) can be harmful to your health. However, in the United States the labels introduced in 1989 contain warnings about alcohol and pregnancy and alcohol and driving, as well as operating machinery. Although the need for warning labels has been discussed in Canada, no serious government-sponsored legislative or regulatory steps have been taken.

It is too soon to have an evaluation of the impact of warning labels on drinking in the United States, however, they might have some effect on rates of fetal effects. Most of the issues on both sides of the question have been discussed elsewhere (Smart, 1990). We do not have many programs for the prevention of alcohol-related birth defects in Canada, although the problem is substantial in some native communities. Some have objected to labels as unnecessary or ineffective, as few women see the bottle from which they drink. However, some do and they might be influenced if they were pregnant. Another problem in Canada is that warnings would have to be in French and English and that might be difficult to do for lack of space. However, in the United States warnings are often on the back side of the bottle, where there is more space.

The effectiveness of warning labels would be greater if combined with government media campaigns and school programs to educate people about the dangers of drinking and pregnancy. Although the Canadian Medical Association and the Standing Committee on National Health and Welfare have recommended alcohol warning labels, there seems to be no political will to introduce them. Canada should have health warning labels on alcoholic beverages and programs to inform the public about the risks of drinking and pregnancy as well as drinking and driving. They could do no harm and might do some good in an area with very few programs and little research.

Mass Media and Health Promotion Programs

Numerous mass media programs about drinking have been tried in Canada. For example, in the 1980s there was the "Dialogue on Drinking" sponsored by the federal government, the "You are Your Own Liquor Control Board" campaign in Ontario, the AWARE program for responsible drinking in Saskatchewan, various programs on moderate drinking sponsored by distillers, and many others. Everyone seems to favor public education about drinking. Unfortunately, many reviews have shown that none of these programs on their own seem able to change people's drinking behavior. The programs can give information and perhaps change attitudes about drinking; however, actual drinking is rarely changed. Nevertheless, Dialogue on Drinking did stimulate discussions about drinking and local initiatives to deal with drinking problems.

Some recent evaluations have been made of anti-alcohol campaigns in Canada. For example, there are evaluations of many anti-drinking and driving campaigns such as the 12-hour licence suspension (Vingilis, Blefgen et al., 1988), random stopping programs (Vingilis and Salutin, 1980), as well as public information campaigns about alcohol risks and driving (Farmer, 1975; Hubert, 1990). Generally, all of these programs have been effective and their collective impact has probably been to greatly reduce impaired driving and alcohol-related accidents over the past 10 to 15 years (Smart and Mann, 1995).

A large number of health promotion programs have been devised in the 1980s to improve people's health and lifestyles. Many of them involve information about better eating habits, avoiding smoking, weight loss and safer alcohol consumption. We have no way to count all of the programs put on by schools, governments and private companies. However, their number has expanded rapidly in the last decade. Macdonald and Wells (1993) surveyed programs at Ontario work sites and found that in 1989 only 35.8 per cent of work sites had a health promotion program, but that by 1993 the proportion had risen to 52 per cent. No doubt these programs have contributed to the decline in alcohol consumption and problems, but detailed research has not been done on this topic.

Much research has found that public education is most effective when combined with a change in laws and in their enforcement (Blane and Hewitt, 1972). For example, several studies have shown that when drinking-driving laws are being heavily enforced, public education is effective. Also, when the .08 BAC law was passed in Canada, people were informed about it through mass media; the risk they saw of themselves being caught probably increased, and there were fewer drinking drivers, at least for a time (Smart, 1972). What is needed now in Canada is a policy to further reduce per capita alcohol consumption by price manipulations, combined with mass media programs to tell people why the policy is necessary and to emphasize how safe alcohol consumption is part of a larger health promotion approach. This program would be far more effective than merely encouraging people to be responsible drinkers, to be their "own liquor control board," or to have a "dialogue about drinking."

Community-based Programs

In community-based prevention programs, the typical focus is everyone at any level of risk in the community, rather than special groups such as students or parents. These programs tend to be large in size, well-funded and extended over several years. A number of such programs have been evaluated and even more are underway, but at too early a stage for evaluation. They represent the "new wave" of prevention as programmers try to move outside the school to engage large numbers of people in preventing alcohol and drug problems. Unfortunately, some community programs have been expensive, as they involve long time frames and a large staff.

One of the earliest and largest community prevention programs ever developed for illicit drug use among adolescents was done by Pentz and her colleagues in Kansas City (Pentz, Dwyer et al., 1989). It had as its aim to decrease alcohol, tobacco and cannabis use among adolescents, their parents and other residents of the city. This was a large-scale study lasting more than five years. The following components were used:

(i) a 10-session school program for training youth to resist drug use;

(ii) ten interviews and role-playing sessions with family members to get them to discourage drug use;

(iii) mass media reports such as newspaper articles and one- to two-minute news clips about the program (about 16 television events, 10 radio and 30 newspaper events) and a televised press conference;

(iv) training of nine television station managers, 65 teachers and four students per class in drug prevention skills;

(v) special public meetings of community leaders where drug problems are discussed.

In the experimental area (Kansas City) after the program, the rate of alcohol use in the past week was 11 per cent versus 16 per cent in the control city with no program. Although the Kansas City community prevention program was a success, the differences created were not large numerically. Of course the program was very costly and time consuming. The cost was reported to be about $17 million. This program required a large staff and a concerted effort over a long time. Probably this type of program is practical only for communities with many resources and a long-term commitment to preventing the problems of alcohol. For smaller communities with more limited professional resources it may not be a useful model.

Only one major community development project on alcohol and drugs has been fully reported in Canada. This program was designed by Giesbrecht and his colleagues (Giesbrecht, Conley, et al. 1990; Giesbrecht, Pranovi and Wood, 1990) as an assessment of the distribution of consumption model developed by Ledermann (1956). This model specifies that a change in the proportion of heavy consumers could reduce overall per capita consumption. An intervention plan was devised for an experimental community and there were two control communities that did not get the interventions. The interventions included: (i) an education and counselling program for heavy drinkers contacted through advertisements and local physicians, (ii) media ads and mailings about heavy drinking; (iii) presentations and workshops for professionals, unions and the general public and (iv) meetings with health and social service committees and local professionals. All of these were organized to promote low risk practices around alcohol consumption. The program increased the visibility of alcohol issues in the mass media and did reduce alcohol consumption among some heavy drinkers. Surveys before and after the interventions indicated that the experimental community reduced its drinking slightly, relative to the other communities. However, in the experimental community there was no greater decline in alcohol consumption, related problems such as hospital admissions or alcohol-related arrests, compared to the control communities.

Many problems arose in the conduct of the project as often happens in community-based initiatives (Giesbrecht, Pranovi and Wood, 1990). Despite many preliminary discussions there was resistance to the interventions for heavy drinkers. Few physicians referred patients and there was active resistance in the treatment community. Perhaps too few heavy drinkers were referred to affect overall consumption much. Also, community structures and power networks were perhaps not fully used as the planning and project staff came from outside the community. If so, it is difficult for project staff to be accepted and avoid the status of "interloper" or "parachutist" in a community that is small and somewhat closed. Of course, few small communities have the resources or people who could put on large-scale projects without outside help.

Several smaller community projects have been described in Canada. For example, Douglas (1990) developed a program to inform the people of Thunder Bay, Ontario about rules against drinking in parks and unlicensed recreational facilities. The campaign started in 1983 and ran for a year. It involved news items, public service announcements, and paid media advertisements. Also, pamphlets, posters and theme buttons were given away. After the campaign residents were more likely to comply with the new rules (compared to a control community), less supportive of underage drinking and drinking in parks, and expressed more intent to rent facilities not serving alcohol. There were also few reports of alcohol problems in recreational facilities. Unfortunately, data on the effects on youthful drinking were not reported. Many youth, especially problem drinkers, do drink in parks and in recreational facilities and it would be useful to know how to help them.

Caverson (1990) has described a "Sober Driver" program that was directed at non-drinking drivers found in spot checks. The program rewarded sober drivers by thanking them and giving them a plastic licence holder. At the same time there was an extensive media program about drinking and driving. There were problems in collecting spot-check data from control areas and in completing a telephone survey about the campaign. Hence conclusions about the impact of the program are very tentative.

In conclusion, community programs for alcohol prevention have yet to show much impact in Canada. The preference for most researchers is to develop community programs in smaller centres, relatively isolated from big-city media and other influences. This means that such programs do not arise from the community but are grafted onto them from the outside. Perhaps grassroots community programs would have a better chance of success.

Some Innovative Alcohol Prevention Programs for Young People

A variety of innovative programs for primary prevention have been developed in Canada, but few have been evaluated extensively. Most of these programs have been directed at young people, but some are for more general populations or for drinking drivers.

A program called "Booze" involved showing a set of short plays or skits about drinking to young people. It was developed originally by staff and students at Confederation College in Thunder Bay, Ontario. The series of skits involved problems of interest to young people and ended with one that emphasizes the conflicting messages we all receive about drinking. The play was shown to improve the attitudes of students in Grades 9 and 10 and to have a modest impact on drinking, compared to other classroom presentations (Gliksman, Douglas and Smythe, 1983). The major impact would almost certainly be on the young people who performed the skits day after day, but apparently this has not been evaluated.

Nova Scotia sponsored two programs with the same effects. The first, "Drug Awareness Weeks," involved a major media strategy including television, newspaper, magazine, and cinema advertising (Charlebois and Poel, 1982). There were also shopping mall displays and mailings of pamphlets to students. Many people became aware of the program and increased their knowledge levels and attitudes in healthy directions. However, there was no effect on people's drinking when a comparison was made before and after the campaign.

A program called "Decide" (Discover an Effective Club in Drug Education) employed peer education concepts. Club members in high schools explored attitudes and practices around drinking and drug use and then learned to deliver messages about alcohol and drugs to younger students. Where the program was used, students were more aware of drug-related issues. Also, "Decide" members were more assured and more self-confident than their peers. However, no convincing effect on actual alcohol use has been demonstrated.

One of the largest prevention programs for young people and substance abuse was mounted by the Alberta Alcohol and Drug Abuse Commission. The campaigns extended over about ten years and a variety of survey results are available. The campaigns included television, radio and print messages encouraging youth to avoid drug use and encouraging safe alcohol use. There was also a free cartoon-based magazine "Zoot Capri" for young people, which featured articles and themes for young people, but cautioning against drug use of all types. The evaluations showed that the campaign was recognized and recalled by most teens and parents. Many reported that elements of the campaign helped them avoid pressures to use drugs. Also, alcohol use declined in Alberta during the early part of the program. However, later evaluations showed that although alcohol use continued to decline, it declined more in Manitoba, which did not have the program. Some data also indicated that Alberta teenagers were drinking at earlier ages than in Manitoba.

Another important mass media persuasion effort was launched by Health and Welfare Canada. There were three campaigns, "Really Me" (discouraging drug use), "Play It Smart" (discouraging drinking and driving) and "Break Free" (aimed at reducing tobacco use). The "Really Me" campaign involved several television commercials, print ads, posters and The National Drug Test — a 60-minute television program. Most Canadian youth were aware of the campaign, were interested in the ads and found them believable. By the end of the campaign somewhat more parents reported conversations with children about drugs.

The "Play It Smart" campaign involved similar activities but was less intensive. Close to 90 per cent of young people surveyed were aware of the campaign and a majority recalled the slogan. Also, about 75 per cent of young people said they would think about drinking and driving, and about 25 per cent would talk to their friends about it.

An interesting program called "Safe Grad" has been devised in Saskatchewan to cope with the large number of alcohol-related deaths accompanying the parties after high school graduations (Kleisinger, 1984). This phenomenon seems to be largely based in Saskatchewan and Alberta, where large unsupervised parties are often held in rural areas. "Safe Grad" puts on seminars for selected students from many boards of education. They get training in how to run a safe graduation program in their school, and the seminar ends with an alcohol-free "Safe Grad" party. Student organizers are advised to provide lots of food and non-alcoholic drinks and to make sure buses and taxis are available for those who have too much to drink. Some evidence shows that alcohol-related accidents have been reduced following "Safe Grad," but the series is short (five years) and control conditions with no program have not been used.

It should be remembered that sometimes alcohol prevention programs backfire and have unintended effects. For example, a program conducted in six Western Ontario communities, entitled "Decisions and Drinking: The Power of Positive Parenting," consisted of structured learning experiences for parents of young children, including training in role-modelling, decision-making and parental alcohol use (Albert, Simpson and Eaglesham, 1983). Quite unexpectedly, the program changed parents' attitudes in an undesirable direction: more favored drinking. However, knowledge levels did increase. Although the authors provided some methodological explanations for the undesirable attitude shift, the study does remind us that alcohol prevention is difficult to do and uncertain in outcome.

Alcohol Education in Schools

Traditionally, school alcohol education has been done with a temperance orientation. In the heyday of the temperance movement, special textbooks on alcohol were made available, and temperance workers often visited schools to talk about "the demon rum." These programs persisted in some provinces into the 1960s, but all seem to be gone now. However, the school has all children captive at the time when they are learning to drink, and it is therefore a good setting for education about alcohol. Alcohol education also fits in with the new trend towards non-traditional subjects and the movement for schools to educate about a variety of social issues such as sexual behavior, values and morals.

Many schools do not have alcohol education, and many teachers are uncomfortable with the topic. Perhaps teachers have had no specific training in how to do alcohol education programs or direct experience with such programs. Of course, many school curricula are packed with compulsory subjects, and time to do all of the extras is often not available. In Ontario schools, a survey in 1977 reported that only 54 per cent of children had any alcohol education at all and many had only one or two classes (Smart, Adlaf and Walsh, 1991). Certainly, a few classes per year will have very little impact. However, there has been some improvement In 1991, 78 per cent of students had alcohol education and the number of classes also increased. We have found a close correlation (r = .95) between the increase in alcohol education and the reduction in drinking levels. As alcohol education increased, drinking decreased greatly and alcohol consumption among students dropped (Smart, Adlaf and Walsh., 1991).

To some extent, alcohol education was not given a chance in the past because few good programs were available. However, that has changed in the past decade. A program entitled "Making Decisions" has been developed in British Columbia for Grade 6 students. Since most students at this level have not begun to drink or use drugs, the program attempts to get them to make wise decisions and avoid peer pressures. There are eight activity-centred lessons. The program is now recommended for Grade 6 students in Alberta and is also being used in Yukon and the Northwest Territories.

A variety of tested, effective programs have been developed by the Addiction Research Foundation in Toronto (Goodstadt and Sheppard, 1983). The clear lesson plans can be easily used by teachers and are well received by students. The ideal lesson plan involves 10 lessons at each of two grade levels, thus greatly increasing the usual exposure of students to alcohol education. The program is now used in many schools in Ontario and in some other provinces.

Goodstadt and his colleagues (1982) examined the effects of these alcohol lesson plans for Grades 7-8 and 9-10. Results were mixed but generally positive:

(i) Both programs increased knowledge levels. Among males attitudes became more favorable to alcohol after the program.

(ii) Effects on alcohol-related attitudes were mixed — attitudes among drinkers became more pro-alcohol, but for non-drinkers they became less pro-alcohol.

(iii) For students in Grades 7 and 8, the program had no effect on expectations to use alcohol less often, however, Grade 9 and 10 students in the program were more likely to expect to use alcohol less in the next year than were controls.

(iv) The size of the effects on expected alcohol use depended on the number of lessons taught. This research shows that alcohol education can have complex, even contradictory effects, and that effects are different for different groups.

A later study by Goodstadt and Sheppard (1983) examined the effects of three approaches: a cognitive program, a decision-making program and a values-clarification program. Students endorsed the cognitive program most strongly. It improved levels of alcohol knowledge, but not substantially, whereas the other programs did not. There were no program differences in attitude change. For the cognitive and decision-making programs there were no improvements in reported alcohol use or expectations to use. However, those in the values clarification program increased their alcohol use. There were problems in presenting decision making and values programs as these require students to learn in a way to which they are unaccustomed. Nevertheless, the results show the inconsistency typical of many program evaluations.

A study by Schlegel and his colleagues (1984) had results that are somewhat similar to those described above. They compared a "facts" presentation to others with values clarification and decision-making components. Those in the "facts" group had lower alcohol consumption after the program, however, those in the other groups showed no improvement.

Another novel approach at the university level involved both persuasion and policy changes. The Campus Alcohol Policy and Education (CAPE) program included a variety of presentations to students (booklets, posters, talks) on drinking appropriately and without problems, and efforts to get university tavern managers to prevent excessive drinking by students. Students exposed to the CAPE program, compared to those not exposed to it, improved their attitudes about alcohol, their intentions to use appropriately, and reduced their reported consumption (Gliksman, Hart et al., 1990).

If alcoholism education could be expanded to be given at a sophisticated level throughout a range of school years, the effects could be very important for future generations. No doubt, improving the amount and quality of alcohol education will be a difficult and long process for both teachers and schools.

Special Programs against Drinking and Driving

Currently, there is a high level of interest in preventing drinking and driving and the resulting accidents and injuries. Several provinces (e.g., Ontario and Alberta) have introduced temporary licence suspension programs for drivers with small amounts of alcohol in their blood when stopped by police. Proposals have been made to revise the *Criminal Code* and to lower the permissible blood alcohol level from 80 to 50 mg/100 mL, or even lower. In addition, citizens' groups such as Mothers Against Drunk Driving (MADD) and People to Reduce Impaired Driving Everywhere (PRIDE) have been formed in several provinces to support stronger penalties for drunk driving and more public awareness of the issue. There seems to be a public will to do something to prevent drunk driving. However, increased penalties have not worked well, even in Scandinavian countries, which have the strictest laws for drinking and driving, and hence greater penalties may not be of much help.

As with many other areas of alcohol control, there are methods of prevention that probably would be effective but are unacceptable politically. Probably it would greatly discourage drinking drivers if police made drivers leaving the parking lots of bars and taverns take breath tests. Mandatory long jail terms and long licence suspensions for first offenders might dissuade some. Also, having more police spot checks to test for drinking drivers, as is done around Christmas, would also discourage the practice. However, it is not clear whether these would be acceptable legally as random stopping programs have received several court challenges. Research shows that the best method for discouraging drinking and driving is to make people believe that they will be caught and punished. Most people who drive will drink and drive at some time, and very few will in fact get caught. The chances of being caught are only about one in 2,000 in Ontario, so people are right about the low risk.

We can increase people's belief in the risk of getting caught by increasing the real risk. Drinking and driving prevention must be a higher priority with police. There is a need in Canada for better police training to detect drinking drivers, for faster processing of arrested cases, and for more portable breath-testing devices. Some police forces do not have enough portable breath-testers and hence cannot properly enforce the existing laws. Police, too, should establish more spot checks to look for impaired drivers. These are often unpopular with police because they are time-consuming and not seen as "real crime work." They are unpopular with the driving public because they are a nuisance and are sometimes seen to infringe on people's civil rights. Once again, a skilfully constructed public education program could make the program both known and accepted as prevention.

Server Training Programs

In the 1980s new programs based on training servers at drinking establishments were developed in Canada and elsewhere. There were several court cases in which staff and owners of pubs were found liable when patrons became drunk and got into accidents. Insurance companies then began to require server training programs before liability insurance would be issued. These programs usually involve training servers to recognize and not serve impaired patrons, and to serve patrons slowly so they do not get impaired. Several provinces now require such programs for servers. They make sense because a large number of drinking-driving accidents involve persons coming from public drinking places.

In general, research in Canada and the United States has shown such programs to be very effective (Saltz, 1987; Holder and Wagenaar, 1994; Gliksman and Single, 1988). They sometimes lead to reductions in blood alcohol levels for patrons and one study has found a reduction in alcohol-related accidents (Holder and Wagenaar, 1994). However, other studies have found very little impact on server behavior (Howard-Pitney, Johnson et al., 1991) or inconsistent results (McKnight, 1987).

Youthful Access to Alcohol and Drinking-Driving Programs

As stated earlier, all provinces have age 19 as their limit for the purchase of alcoholic beverages except Manitoba, Quebec and Alberta where it is 18. In Ontario, but not other provinces, possession or drinking of alcoholic beverages is also illegal under 19 years of age. Of course, youth over the age of 19 have legal access to alcohol and thus tend to drink more often and have more problems.

Current laws in most provinces require that young people show proof of age or an "age-of-majority" card in order to buy alcoholic beverages. Of course, these cards can be borrowed or obtained illegally. Also, enforcement of the laws for age of majority is weak.

In Ontario the drinking age was increased to 19 in 1979 and young people were required to have age-of-majority cards with pictures to purchase alcohol. In 1981, we included questions on age-of-majority cards in our Ontario student survey (Smart and Adlaf, 1987). Almost no students had the cards (13.9 per cent) but most (75.3 per cent) were drinkers and 99 per cent were underage. Only 9.3 per cent had their own cards, while 4.5 per cent used or borrowed other people's cards as needed. Those who had no card were just as likely to drink in taverns, bars, or restaurants than those who had their own card. However, those who borrowed a card were more likely to drink in pubs or taverns, in cars and at school dances. They drank much more often and reported more alcohol problems than those with no card or their own card. Also, 88 per cent of those with their own card were actually underage. Probably age-of-majority cards are a poor control device for young people. Again, it is a matter of enforcement, but good monitoring of restaurants and bars for underage drinking violations would be a very large task. Of course, it is the job of parents to control drinking of young people when they are at home or in other social or party situations.

Some people have suggested that learning to drink and learning to drive should not be done at the same age, i.e., around 16 to 18. However, changing that would mean increasing the driving age to 20 or 21 since governments don't have much control over the age when people drink. Parents and teachers could be encouraged to get students to postpone drinking; ages of first drinking are rising and perhaps could be made to rise further. Parents could refuse to allow adolescents to be licensed until 21 or so. Governments could raise the driving age to 21 but that would be unpopular, especially with voters aged 18 to 21. The best and most practical policy for both parents and teachers who teach alcohol education is to get students to postpone drinking or at least heavy drinking. However, Stewart and Klitzner (1990) found that contracts with parents not to drink and drive were ineffective.

Some countries have experimented with special laws for young people such as a lower BAC (Blood Alcohol Content) limit. An evaluation study in Australia (Drummond, Cave and Healy, 1987) found that zero BAC limits reduced adolescent nighttime, weekend driving, which is a time for many alcohol-related accidents. The state of Maine has a BAC limit of 20 mg/100 mL and a one-year licence revocation as a penalty for youth who drink and drive (vs a 100 mg/100 mL limit for other drivers). Initially, Hingson and his colleagues (1986) found that self-reported drunk driving and accidents declined as did actual personal injury accidents, but later results were less striking. Also, some 18 U. S. states have nighttime curfew restrictions for young drivers preventing driving after 10 p.m. or midnight. A study of restrictions in four states found that they dramatically reduced accidents during those hours that were restricted (Preusser, Williams and Zador, 1984). Recently laws have been changed in Ontario so that there is a lower BAC for young drivers and they are no longer allowed to drive at night. Both lower BACs and driving curfews for youth should have beneficial effects on alcohol-related accidents. Expansion of such programs to other provinces is warranted.

Another method for reducing youthful drinking-driving risks involves education or special courses. Most young people learn to drive through driver education courses and many insurance companies require them or offer lower rates to graduates of such courses. Some criticism has been made of driver education merely because it increases licensing of young drivers who would not otherwise be able to get a licence (Stewart and Klitzner, 1990).

Mann and his colleagues (1986) conducted a study of school-based drinking and driving programs. Many of those programs were for people convicted of impaired driving, few of whom were young people. However, some were driver education courses for young drivers. Most programs were able to show short-term improvements in knowledge and attitudes about drinking driving. However, few have examined impacts on actual drinking driving or alcohol-related accidents.

One of the few Canadian studies (Albert and Simpson, 1985) evaluated the impact of a drinking-driving program based on the Health Belief Model, i.e., the concept that people mostly evaluate options and choose the one that is most likely to maintain or increase health. The program was given to Grade 11 students and a control group was also included. Those who took part in the program improved their knowledge and attitudes and had decreased intentions to drink and drive. However, those in the experimental group reported an increased level of drinking. It is not clear whether this was an experimental effect or due to non-random assignment to groups. On the basis of current research it is not known whether drinking-driving education can reduce alcohol-related accidents. It may have untoward effects and a strong policy shift toward such programs is not warranted.

Secondary Prevention: Early Detection

Secondary prevention involves identifying people with minor alcohol problems, or problems at an early stage, and offering treatment or some other intervention that prevents the problems from getting worse. Some studies have found that younger alcoholics are easier to treat, but others have found essentially no difference in outcomes between older and younger alcoholics. Also, many alcohol problems among young people disappear with age and without any intervention being needed. Nevertheless, chronic alcohol dependence does take a long time to develop — usually 10 to 20 years. Most clinics and hospitals find that many of their clients are aged 35 to 45 and have been drinking excessively for a long time. If they had entered treatment earlier and had shortened their drinking careers, they might have avoided job loss, family disruption, or other life problems.

Employee Assistance Programs (EAPs) represent an attempt to get heavy drinkers into treatment at an early stage — at least while they are still employed (Shain and Groenveld, 1980). There are many different sorts of programs, but most seem to involve the concept that employees with drinking problems are required by the employer to submit to clinical treatment or other interventions to change their drinking habits. The employer agrees to maintain the employee's job and to give the employee some support in arranging treatment, such as time off work. The survey by Macdonald and Wells (1993) showed that 32 per cent of worksites had EAPs in 1993 compared to only 16.1 per cent in 1991. Clearly there is a need for expansion of such programs, especially to smaller companies.

EAPs have had many successes and are generally very useful in getting into treatment heavy drinkers who might not otherwise get there. However, it is clear that they do not identify problem drinkers as early as had been hoped. Several studies have shown that those taken into EAPs have about the same characteristics as those who come into treatment by other means.

Many efforts are being made to improve the ability of physicians to identify alcohol problems at an early stage. Physicians frequently do not identify problem drinkers among their patients. Sometimes they rely on false information about the patient's drinking, or misinterpret early signs of heavy drinking (e.g., fractures, peptic ulcer, sleeping difficulties). At other times, physicians are reluctant to force the issue of drinking as the cause of physical illness when the patient denies that he or she has drinking problems. The Addiction Research Foundation developed a composite set of indicators to identify problem drinkers at an early stage (Skinner, Holt and Israel, 1981). It involves questions about drinking, drinking problems, and medical complications, as well as a variety of newly established blood tests. The use of this composite index should allow physicians to better identify early-stage problem drinkers. At present, more research is needed to determine how successful the index is in identifying problem drinkers and helping to get them into treatment early.

Several new programs involving physicians have been developed in Canada. In one project physicians give a questionnaire to patients who seem to have alcohol problems and some get liver function tests. They are given their results and monitored for some time. Preliminary results seem positive but the study is still incomplete. Another project sponsored by the Brewer's Association Canada and the College of Family Physicians (1994) involves training physicians to recognize drinking problems and to discuss safe drinking levels with patients at high risk.

Tertiary Prevention or Treatment

Issues around treatment for alcoholism or problem drinking are discussed in Chapter 11. We are concerned here only about treatment as a preventive effort — a role in which it is not usually seen. Successful treatment of drinking problems, even at a late stage, may prevent further problems. For example, we know that early treatment of liver cirrhosis (a late-stage complication) can save the life of the alcoholic. Also, successful treatment may prevent suicide and other very late-stage complications such as a progression to skid row. Usually, alcoholics who come into treatment have not hit bottom in terms of having lost everything.

Unfortunately, many studies have shown that some treatments for alcoholism are not very effective (see Chapter 11). We offer treatment for humanitarian reasons and in order to do research on better methods. However, treatment is sometimes effective and hence prevents problems from worsening. Also, much treatment is done by community-based groups such as Alcoholics Anonymous. We have no good evidence to show that their treatment is superior to clinical treatment, but it is certainly cheaper for the taxpayer.

Recent research has clearly shown (Smart and Mann, 1995) that increased levels of treatment and Alcoholics Anonymous membership are associated with declines in liver cirrhosis and other problems. Almost certainly the connection is more than accidental. When large numbers of alcoholics are treated, some improve, and only small reductions in drinking are needed to reduce liver-cirrhosis risks. Planning for prevention of alcohol problems should also, therefore, include improving access to treatment for those who already have alcohol problems.

The Need for Balance between Treatment and Prevention

The dominant priority in the alcohol field has always been treatment. Both primary and secondary prevention have been relatively unimportant in most provinces and have not attracted the needed funding. Many programs have nothing but treatment facilities, and in others treatment predominates. In most cases, treatment for alcoholism has not been shown to be much better than no treatment at all. However, there is considerable public and political support for treatment; people want to have facilities providing care for alcoholics. There is, at the same time, a new recognition in Canada of the need for primary prevention. It is now realized that alcohol problems are essentially "lifestyle" diseases and that they are largely self-inflicted. With this has come the recognition that primary prevention is not only possible but essential. There is a clear need to re-establish a balance between primary prevention and treatment, and to use all available efforts to prevent new cases.

Summary

Prevention of alcohol problems has never been as popular as treatment in Canada. Most areas have treatment programs but no real prevention activities. The concern should not be the prevention of all drinking, but only drinking that results in harm to the user or society.

Some models of prevention dictate that heavy drinkers should be the target; others that social attitudes should be modified. However, most evidence seems to support the view that governments should reduce per capita alcohol consumption, probably by raising the price of alcoholic beverages. In order to be successful, such an approach (and any other) would require public support.

Canadian prevention programs have involved changes in government policies and special programs to change attitudes and behavior around drinking. There have also been many successful prevention programs, especially for young, drinking drivers. For example, programs getting students to postpone drinking, lowering BAC limits for young drivers and some driver education programs.

There are also successful EAPs that get problem drinkers into treatment at an early stage. We need an expansion in these programs as few industries have them yet. Also, some early identification programs are very promising, especially those run by physicians. Most of these are relatively cheap and hence have a high cost-benefit ratio. Some shift in the balance between prevention and treatment towards more prevention has taken place since the 1980s and we need to continue this development. In order to be effective, prevention programs should involve changes in the drinking age, changes in licensing laws, and mass media campaigns aimed at changing public attitudes towards drinking and driving and dangerous drinking in general. Research has also shown that large increases in treatment levels and educational programs are associated with declines in alcohol-related problems. In almost all communities there is a need to expand such programs.

Chapter 11

Treatment For Alcohol Problems in Canada

Until sometime around the end of the Second World War, chronic, excessive drinking was generally regarded as a symptom of moral weakness rather than a condition requiring state supported treatment. Although many temperance workers and prohibitionists were aware of the disease model of alcoholism, they did not lobby much for treatment programs until the 1940s. Rather many focused their attentions on the need for social and legal controls on drinking behavior. Temperance societies encouraged inveterate drinkers to sign pledges of abstinence and welcomed them to speak about the evils of drink at meetings. However, treatments for alcoholism were not always priorities within the temperance or prohibition movements during the 19th and early 20th centuries (Rush and Ogborne, 1992).

Toward the end of the 19th century, a small number of private sanatoriums for the treatment of "inebriety" and "narcomania" were established by physicians with the support of wealthy businessmen. One of the earliest was the Homewood Retreat in Ontario (Warsh, 1989). This was established in 1882 and its first medical director, Dr. Steven Lett, became an outspoken proponent of treatment for alcoholism and drug addiction. Treatments provided at Homewood included the administration of tonics and bromides, "gold cures," water-bath therapy, rest and recreation. Deer Park Sanatorium, established in Toronto in 1891, was another influential facility that used similar treatments.

Although based on the notion that alcoholism is a disease, the main focus of treatment at private sanatoriums was the management of withdrawal symptoms. The avoidance of these symptoms was assumed to be the main reason for drinking among alcoholics but relapse after abstinence was still considered a moral issue. Thus the adoption of Christian values and lifestyles were considered critical to long-term success. At Deer Park "one of the main objectives is to surround a patient with a Christian influence and send him out, not only cured of his appetite for alcohol and other narcotics, but a restored man" (*Quarterly Journal of Inebriety*, 1892).

Private sanatoriums generally catered to the middle classes although some also admitted less-well-off clients as charity cases. However, the overall proportion of problem drinkers treated in sanatoriums was certainly very small. A few others were treated in mental hospitals, and, as the century progressed, in programs run by the Salvation Army. In some provinces, ministers of health were required to provide facilities for the treatment of alcoholics committed for treatment by the courts. However, until the end of the Second World War, most Canadian governments provided few resources for the treatment of alcoholism and the condition was largely ignored by all but a few in the medical profession.

From the 1940s to the 1960s

Many new alcoholism treatment services were established during the 1940s and 1950s, partly in response to the lobbying efforts of members of Alcoholics Anonymous (AA). This remarkable and unique movement was founded in 1935 in Akron, Ohio by two chronic alcoholics, one a stockbroker (Bill W.) and the other a physician (Dr. Bob). From the beginning, AA promoted the view that alcoholism, although incurable, could be successful arrested if treatment was provided for withdrawal and alcoholics followed a 12-step program of recovery. The movement spread quite rapidly and the first Canadian AA meetings were held in 1943. With the support of some physicians and community leaders, Canadian AA members lobbied governments to provide funds for treatment and education programs.

Efforts to secure government support for alcoholism services were also spurred by changing views of the nature of the addiction. Following Prohibition, the view of alcohol as an inherently addictive poison was tempered by the notion that alcoholism does not simply result from excessive drinking but from a combination of factors including individual susceptibility. Further, the notion that continued drinking in the face of difficulties and relapse after abstinence were signs of moral weakness was replaced by the view that these behaviors were symptoms of a condition (sometimes called addiction or dependence) for which the drinker was not responsible. These new conceptions of alcoholism helped to persuade governments that alcoholism could be prevented, not simply through regulations on drinking behavior but through education and other means. Dr. Gordon Bell, founder of Toronto's Donwood Institute, was particularly active in the promotion of this new conception of alcoholism (Bell and Soloman, 1989).

Dr. E.M. Jellinek, an American physician working at Yale, also had a great deal of influence on the postwar development of Canadian alcoholism treatment systems. His book *The Disease Concept of Alcoholism* (Jellinek, 1960) was highly acclaimed and he also worked for a term at the Addiction Research Foundation in Toronto. Further, many Canadians involved in the development of alcoholism services met Jellinek and his colleagues at Yale summer schools and at other times.

It is, however, important to recognize that Canadian governments and most Canadian treatment professionals have never fully subscribed to a simple disease notion of alcoholism. To this day, that nature of "alcoholism" continues to be hotly debated among researchers and policy makers and many doubt that "alcoholism" is a distinctive phenomena or even a useful concept (Fingarette, 1988). In contrast with the United States, where the disease model of alcoholism seems to be well entrenched and promoted by celebrity "alcoholics" in books and on television talk shows, Canadians seem less willing to accept simple categorizations and explanations. As we will see in this chapter and in Chapter 12, Canadian efforts to treat and prevent excessive drinking reflect an appreciation of its complex causes and the need for multiple approaches.

By the end of the 1950s most provinces had established alcoholism foundations, commissions or departments with responsibilities for treatment, prevention and, in some cases, research. The first provincial alcoholism agency, the Addiction Research Foundation (ARF),[1] was established in 1949 in Toronto through the efforts of H. David Archibald and some temperance workers in politics. Archibald became its first executive director and has written a highly readable account of his time at the Foundation (Archibald, 1990).

From the Late 1960s to the 1980s

Canadians witnessed a remarkable growth in the number of addiction treatment programs during a 15-year period starting in the mid-1960s. Thus of the 340 or so specialized programs operating in Canada in 1976, two-thirds had been established since 1970 (Reid, 1978). Most of these programs were primarily for alcoholics, although many also served people who used both alcohol and other drugs. Many different types of services were established including detoxification centres, outpatient services and long-term residential services. New services were established both in provinces with foundations or commissions and in others where commissions has been absorbed by government departments.

1. Originally the Alcoholism Research Foundation.

During this period there was also a broadening of the range of professionals involved with the treatment system. Initially, medical and paramedical professionals predominated in treatment agencies. However, during the 1960s and beyond, psychologists, social workers and mental health counsellors became more prominent. Some treatment agencies have also always employed self-identified recovered (or recovering) alcoholics as counsellors, especially detoxification centres and some residential treatment centres. Many of these are competent and dedicated people who have also received certification from agencies that promote of high standards of professionalism. The certification process typically involves a review of experiences and qualifications, letters of recommendation, an interview and sometimes a written examination. Further details of the development of alcoholism services are given by Rush and Ogborne (1992).

From the 1980s to Today

The rapid expansion in the number of addiction services that occurred in the 1960s and 1970s has, for the most part, been followed by a period of stabilization and consolidation. Some addiction services have recently been established or expanded through special provincial or federal initiatives, especially services for native people, youth and those in the correctional system. However, significant increases in government spending on addiction services are very unlikely in the short term. In the 1980s and 1990s Canadian governments have been increasingly concerned with controlling expenditures on health and social services, and some cutbacks in spending on these services have taken place. Further cutbacks are also anticipated. Although addiction treatment services have not been significant victims of recent funding cuts, many service providers are dealing with flat-lined or decreasing budgets.

The current supply of specialized addiction treatment services can only be estimated because there is no national database and different regions keep records in different ways. However, on the basis of entries in treatment service directories, there seem to be more than 1,000 publicly funded addiction treatment agencies across the country. Many of these have branch offices, so the actual number of physical locations where addiction treatment is provided is much higher. Addictions counselling and other forms of treatment are also provided in general and psychiatric hospitals and a variety of other settings by physicians and psychiatrists, social workers, psychologists, probation officers, school guidance counsellors, ministers of religion and others with influence. Many people also join special self-help groups such as

Alcoholics Anonymous or Women for Sobriety. However, little is known about the extent and nature of treatment provided outside of specialized services. Details of specialized services, their clients and the treatments provided are also unavailable for the country as a whole.

Details of specialized addiction treatment services in Ontario are available through surveys conducted by the Addiction Research Foundation. The latest survey (Ellis and Rush, 1993) showed that Ontario had 229 specialized addiction treatment agencies. Most of these serve people with alcohol or with alcohol and other drug problems, while very few serve only people using drugs other than alcohol. Ten per-cent of Ontario's services were hospital-based detoxification services, 16 per cent provided short-term residential treatment, 29 per cent provided long-term residential treatment and support, 29 per cent were designated as non-residential services and 17 per cent were described as assessment/referral services. Many of these also provided outpatient counselling. The operating cost of Ontario's addiction services was estimated to be about $100 million, of which $92 million (84 per cent) was provided by the provincial government, mainly through the Ministry of Health (80 per cent of all funds). Other significant sources of funds were the provincial ministries of Community and Social Services and Corrections (11 per cent and 3 per cent of all funds respectively) and the federal government (4.3 per cent of all funds). No comparable figures are available for Canada as a whole.

Most of the direct costs of specialized services are borne by the provincial governments, even in Quebec where there are a significant number of for-profit treatment services. However, the federal government funds most treatment programs for native people, residents of the Yukon and the North West Territories, the RCMP, federal prisoners, members of the armed forces and war veterans. In the late 1980s and early 1990s, some federal funds for treatment of women and youth were also made available under Canada's Drug Strategy. These could be used by provinces willing to contribute 50 per cent of the costs of new or expanded service. The total cost of specialized addiction services for Canada as a whole is unknown.

Because alcoholism treatment is largely a provincial matter, there are some important regional differences. However, it can be said with some certainty that across the country there is an increasing emphasis on non-medical, community-based services and the need for early intervention. This reflects the impact of research supporting the cost-effectiveness of outpatient treatment. However, decreasing health care budgets are also influencing the delivery of alcoholism services and non-medical, community-based services are clearly less expensive than hospital-based, medical services. The increasing emphasis on community-based services is also a reflection of a growing appreciation that alcohol abuse is a complex, partly learned, behavior that is influenced by social conditions. Simple "disease" models are unable to account for observable realities and programs that fail to address these behavioral and social factors have not proven very effective (Miller, Brown et al., 1994).

There are signs of a growing interest in the use of court-ordered or mandated treatments for people with alcohol and other drug problems. For example, a federal bill (Bill C-41) that makes extensive provisions for court-mandated treatment was expected to come into force in 1996. Currently such treatments occur to a varying extent in all parts of the country and especially for people convicted of second or subsequent drinking and driving offences. However, there are wide inter- and intra-provincial differences in the ways in which the justice system responds to people with alcohol problems. Decisions to order treatment seem to depend on a mix of factors, including offender preferences and the opinions of lawyers, judges and probation officers. Of course, the availability of suitable treatment programs also influences court decisions.

Little is known about the actual prevalence of mandated cases in treatment and this is likely to vary with the characteristics of clients. In Ontario, 15 per cent of referrals to treatment settings other than detoxification centres are from police, the courts or probation officers (Ellis and Rush, 1993). However, it is not known how many of those referred from the criminal justice system were seeking to impress the courts or were required to seek treatment as a condition of probation.

Under the *Mental Health Act*, it is also possible for physicians to arrange for alcoholics to be detained in order to receive treatment. However, this occurs in very few cases.

Who Seeks Treatment?

Most people who experience alcohol-related problems do not turn to specialized addiction treatment services for help. Of those who do, the majority use self-help groups. This is evident from general population surveys in which people identified as having drinking problems are asked about their use of treatment services. The most recent Canada-wide survey in which people were asked about treatment for alcohol problems was conducted in 1989 and involved telephone interviews with over 11,000 people. Of those who reported that drinking had a negative effect in any of six life areas, only 8.5 per cent had sought help for alcohol problems (Rush and Tyas, in press). Of those who had sought help, only 12 per cent had used specialized treatment agencies while 77 per cent had been to Alcoholics Anonymous. These figures probably under-estimate the true rates of treatment use because heavy drinkers and people in residential treatment are usually under-represented in population surveys. Also, the figures do not tell us what proportion of problem drinkers will seek help over their life-times. However, the results of this survey are generally consistent with those of other Canadian, American and British studies. These all show that most problem drinkers do not use specialized treatment agencies.

The 1989 Canada-wide survey also showed that, consistent with other studies, a substantial number of those who had sought help for alcohol problems did so from more than one source. Thus 16 per cent of help-seekers had used two services, 5.4 per cent had used three services and 2.8 per cent had use four or more services. Male problem drinkers were more likely to seek help from a specialized service than females. Compared with those who do not seek help, those seeking help tend to be older, to have lower incomes, to be less likely to describe themselves as "students" or as "keeping house," and to be more likely to speak English or French at home.

Help-seeking for drinking problems often occurs at the suggestion or insistence of others, and Canadians are quite likely to suggest some form of help to family members, friends and colleagues who have drinking problems. This is also clear from population surveys. For example, in the 1989 Canada-wide survey, 68 per cent of respondents said they would suggest treatment to a partner who had alcohol problems. The majority of these said that AA would be the treatment they would recommend, while only 8 per cent said they would recommend help from a specialized addiction treatment agency.

Studies of people in treatment programs also show that external pressures often influence the decision to seek help. These include pressures from spouses and employers, and pressures arising from brushes with the law. Alcohol-related health problems also motivate many people to seek help. In short, as has often been said, people seek help for alcohol problems because of the four "L's" — lovers, livers, livelihoods or the law!

Over the past few decades, the average age of the population in specialized addiction treatment service has declined (Rush and Ogborne, 1992). The average is age is now between 30 and 35 and many more teenagers and young adults are presenting for treatment than in the past. Treatment systems have also witnessed an increase in the number of women seeking help. These changes partly reflect changes in drinking habits. However, they also reflect closer links between addiction services and other health and social services agencies. Changing attitudes to treatment and efforts to involve people in treatment before they become severely alcohol dependent have also contributed to changes in the treatment-seeking population.

Traditional "skid-row" types of drinkers are becoming increasingly rare in most parts of the addiction treatment system — even in detoxification centres, which were originally designed for this kind of client. This reflects a decrease in the size of the skid-row population and the efforts of treatment services to broaden their scope. Skid-row drinkers have not attracted much attention lately and it is not clear why there are fewer such people than in the past. The success of prevention and treatment programs may be part of the explanation but it is more likely that the former generation of skid-row men are dying off and not being replaced by people with similar needs and experiences. Many traditional skid-row types were war veterans with life experiences that were quite different from those of subsequent generations. There is, however, a growing concern with a new generation of socially unstable young people and fears that they may eventually come to resemble old-style skid-row types. These so-called "street youth" tend to drink heavily and to use many drugs. They appear in youth-oriented treatment services and some special programs have been established for the street youth population. However, they have many problems and tend to drop out of treatment (Smart and Ogborne, 1994).

Since the 1960s and the rise of the "drug problem," an increasing number of people have developed problems with both alcohol and other drugs. Some of these are "traditional alcoholics" who medicate themselves with drugs. Some are primarily drug users who use alcohol when drugs are not available. Others are true "poly-drug" users, who use a variety of chemicals indiscriminately or with some understanding of drug interactions. Multiple drug users are now frequently seen in addiction treatment programs including programs that have traditionally served only alcoholics (Rush and Ogborne, 1992). Their presence is leading to changes in ideas about addiction and to the development of new treatments.

There have been few recent, detailed studies of populations in treatment and we should be cautious about generalizing from older studies given recent trends. However, it is probably fair to assume that different programs continue to serve different populations in much the same way as in the past. Thus outpatient services see a wide range of clients with varying degrees of dependency, while community- and hospital-based residential programs see people with more serious problems. Halfway houses tend to serve socially unstable people with a long history of alcohol-related problems.

What Kinds of Treatment are Used in Canada?

Treatments for alcoholism have one or both of two main goals. Either they aim to help the drinker withdraw from alcohol or to prevent a relapse to heavy drinking. Typically relapse-prevention treatments also aim to help the drinker to resolve problems in other life areas.

Treatments aimed at assisting drinkers to withdraw from alcohol may be provided in residential detoxification centres located in hospitals or in the community, in other hospital beds, in other residential settings or on an outpatient basis. Non-medical, residential detoxification centres are found in several provinces and especially in Ontario, which has 35 such centres. These are administratively linked with hospitals but are otherwise non-medical, in that they do not provide medication to alleviate withdrawal symptoms. In lieu of medications, non-medical detoxification centres offer social support and encouragement, vitamins and other nutrition during withdrawal. However, people with very severe withdrawal symptoms (less than five per cent of admissions) are referred to the sponsoring hospital (Gavin, Bohm et al., 1987). Counselling and referral are offered once the withdrawal is complete. In other provinces (e.g., British Columbia) detoxification centres involve more medical and para-medical staff and make more use of medications to alleviate withdrawal distress. Medications to alleviate withdrawal distress and to prevent delirium tremens may be provided in hospital and medical outpatient settings. The most commonly used medications are benzodiazepines, especially Valium.

Treatments designed to prevent relapse to heavy drinking following withdrawal are provided in a variety of non-residential and residential settings. These include the offices of general health and social service providers and specialized outpatient or day addiction centres. Some general and some psychiatric hospitals have specialized addiction units and there are specialized short- or long-term residential programs in many communities. Special addiction programs are also offered in many prisons and other correctional centres.

Treatments provided by non-specialists generally involve some form of individual or group counselling. General practitioners, other physicians and psychiatrists may also prescribe tranquillizers or other drugs to treat underlying problems such as anxiety or depression. Medically qualified practitioners may also prescribe the so-called "anti-alcohol" drugs disulfiram (Antabuse), calcium carbide (Temposil or Abstem) and metronidazole (Flagyl). These drugs differ mainly in their duration of action, but they all slow the metabolism of alcohol by the liver, resulting in a build-up of acetaldehyde. As a result, the drinker experiences an unpleasant reaction characterized by flushing of the face, chest pains, pounding of the heart, nausea, vomiting, headache, dizziness, weakness, difficulty breathing and loss of blood pressure. However, the objective of prescribing anti-alcohol drugs is not to make the drinker feel ill when he or she drinks but to encourage the would-be drinker to remain abstinent and thus avoid negative reactions caused by drinking in the presence of an anti-alcohol drug. It is not known how frequently these anti-alcohol drugs are used, either in general practice or in specialized addiction treatment programs in different parts of the country. In Ontario 14 per cent of all treatment programs and 24 per cent of short-term residential programs report using anti-alcohol drugs (Ellis and Rush, 1993). The implanting of slow-release anti-alcohol drugs under the skin has been tried in Canada. However, the effectiveness of this methods has not been proven and it is not currently widespread.

Acupuncture is available from private practitioners in various parts of Canada and it has been promoted as offering an effective "cure." However, most specialists remain unconvinced of its efficacy and probably very few drinkers are treated by this method (McLellan, Grossman et al., 1995).

Electrical or drug-aversion therapies that involve the pairing of drinking with electric shock or with drugs that make the user feel sick are not widely used in Canada, if at all. Some U.S. programs use electrical aversion and others may still use video-taped feedback. This involves taping people when they are intoxicated and later showing them the tapes. This has not proven effective and was probably never used in Canada (Feinstein and Tamerin, 1972). Hypnosis is also an uncommon treatment for alcohol-problems in Canada although practitioners of cognitive therapies may use techniques of suggestion (see below).

Most drinkers who seek help receive some form of individual and group counselling or therapy from either non-specialist or specialist service providers. A wide range of different activities are encompassed by these terms. In general, "counselling" refers to group or individual work that is more short-term, directive, and focused on problem solving and behavioral change. In contrast, "therapy" is used to describe more long-term individual or group work that focuses on the supposed underlying causes of drinking problems. Many "therapists" have been influenced by the ideas of Sigmund Freud or Carl Jung and their associates and followers. However, there are many exceptions to this simple distinction and the so-called "behavior therapies" are very focused on drinking behaviors, the conditions in which these occur and their consequences. "Deep" causes of drinking such "oral fixations" or "unresolved oedipal conflicts" are not a primary concern for behavior therapists although, where appropriate, drinkers would be treated for significant psychological problems especially depression, phobias or anxiety.

Useful descriptions of different types of "talk" treatments were developed for a recent survey conducted in Ontario (Rush and Ellis, 1993). These are summarized here, together with the percentage of Ontario treatment services (excluding detoxification and assessment/referral services) using each method.

> **Psychotherapy** involves individual or group activities. It assumes that alcohol problems are symptoms of underlying psychological problems and seeks to resolve these problems through the use of interpretations and the development of insight. This contrasts with more present-oriented, problem-solving approaches. Fifty-six per cent of Ontario's treatment programs report using psychotherapy.
>
> **Problem-Solving Counselling** involves individual or group activities that have a pragmatic, present-oriented focus on problems of living and practical ways to change problem behaviors. This approach is used by 99 per cent of Ontario's treatment programs.
>
> **Confrontation** refers to intensive, challenging feedback on behaviors and attitudes provided by an individual or a group intending to motivate change. Confrontation is used by 56 per cent of Ontario's programs.
>
> **Family/Marital Therapy** involves the spouse or other family members in the treatment in order to improve communication, problem-solving and other family skills. Fifty-five per cent of Ontario's treatment programs employ this approach.

Contingency Management requires clients develop written contracts in which they agree to access or withdraw from naturally available "rewards" that depend on drinking. For example, clients may agree to put away money that might otherwise be use for alcohol and to buy themselves a gift if they are abstinent for a given period. They may also agree to send money to their least-favored cause if they drink. This approach is used by 47 per cent of Ontario's programs.

Skills Training includes a variety of specific treatment techniques. Skills training approaches, which are used in almost all of Ontario's treatment services, include:

Problem-Solving activities aimed at improving the abilities to analyse and solve problems in daily life;

Vocational Training activities aimed at finding employment or increasing work-related skills;

Leisure Skills Training activities aimed at improving the ability to use leisure time in a more productive and enjoyable manner;

Relaxation/Stress Management activities aimed at improving the ability to relax and cope with stress, including relaxation training, biofeedback, meditation and related techniques;

Behavioral Self-Control aimed at teaching ways of modifying the use of alcohol, often with a moderate drinking goal. Among other things, the client may be taught how to deal with high-risk situations through direct instruction, role playing or the use of self-help books;

Relapse Prevention, which refers to specific techniques designed to teach clients how to cope with high-risk situations and to increase their confidence in dealing with these situations.

Many specialized addiction agencies offer a variety of specific treatments and some can be characterized as "broad spectrum" programs. A typical broad spectrum program would routinely provide alcohol education, relaxation therapy, assertiveness training, leisure counselling, and group and individual therapy and referral to Alcoholics Anonymous. Many programs also provide nutritional counselling that aims to help clients to develop healthy eating habits. Residential programs also sometimes have daily exercise programs based on the "healthy body/healthy mind" principle.

Programs for Special Groups

Agencies that serve only women or youth have been established in some provinces and, as noted in Chapter 6, there are some special agencies for native people. Special addiction treatment programs also exist within federal and provincial correctional systems and, of course, these serve only people in trouble with the law. Some agencies mainly serve socially unstable, "skid-row" alcoholics but the majority of Canadian alcohol treatment agencies serve a wide range of clients. However, many agencies do have special programs for women and young people. A few agencies also have special programs for people who are HIV-positive.

Agencies and programs specifically for women seek to address women's special needs and problems. Many women with alcohol problems have been abused by drunken males and may thus be reluctant to discuss their problems in male-dominated settings. Issues of female sexuality are also best dealt with in women-only programs. In Chapter 5, we noted that women's drinking patterns and motivations tend to differ from those of men and, of course, women are more likely than men to have child-care responsibilities and low incomes. These are further reasons for the development of women's programs.

Young people with alcohol problems are generally less alcohol dependent than older chronic heavy drinkers. However, heavy-drinking youth are likely to also use other drugs and, of course, to have problems peculiar to their age group. These problems include relationships with parents and peers, identity and sexuality problems and problems in school or in finding a first job. For these reasons, many programs targeting youth have been developed, especially non-residential programs that allow young clients to remain in the community while receiving treatment. Supportive residential and outreach programs for socially unstable street youth have also been developed in some communities.

Programs for down-and-out drinkers include those that provide hot meals, clothes and sleeping bags to those who continue to live and drink on the streets, "wet" and "dry" shelters and long-term "dry" supportive residences. Vocational and other rehabilitative services are also provided where appropriate.

Some programs that target people convicted of driving while impaired have also been developed. Some of these use an educational approach and require attendance at a series of lectures or discussion groups. Others are more like treatment programs and focus on individual problems and needs.

Does Treatment Work?

This seemingly straightforward question cannot be answered with a simple yes or no. Much depends on the person being treated, on his or her social situation and, to a lesser extent, on the type of treatment considered (Moos, Finney and Cronkite, 1990). Certainly, many people reduce their drinking after treatment but it is often difficult to disentangle the effects of treatment itself from other influences such as changes in life circumstances, help from friends and families, motivation and other characteristics of the person treated. The fact that many people will solve their drinking problems on their own and without formal treatment is well established (Sobell, Sobell et al., 1993) and indicates the need for caution against claims for the effectiveness of treatment. "Natural" recovery rates similar to those of some treatment programs have been reported, and many treatments in current use have not been shown to produce better results than no treatment at all.

Many questions about the effectiveness of specific treatments remain unanswered and much research remains to be done. At present, the best information available on the effectiveness of treatment comes from experimental studies published in scientific journals. These are studies in which different treatments are compared with each other or with no-treatment controls. Other types of studies, especially those that do not have control or comparison group designs, cannot be used to support claims of treatment success. Many of the experimental studies reported in international scientific journals have been conducted in Canada. As the international treatment research literature has been reviewed several times (Eliany and Rush, 1992; Miller and Hester, 1986a; Holder, Longabaugh et al., 1991), what follows draws heavily on previous reviews. However, we will mention some important Canadian studies and researchers.

Drug Treatments

The tranquillizer diazepam (Valium) has been found to be helpful for the management of severe alcohol withdrawal (Sellers and Naranjo, 1983) and there is also some evidence that certain antidepressants and some new drugs that affect brain cells may reduce craving for alcohol (Anton, 1994). There is very little evidence that the routine use of existing psychoactive drugs reduces drinking and related problems. In fact, there are justifiable concerns that alcoholics who are given psychoactive drugs will become doubly dependent.

Several controlled studies using the "anti-alcohol" drug disulfiram (Antabuse) failed to show any benefits to its use. One Canadian study (Peachy, Annis et al., 1989) showed that patients given the shorter-acting calcium carbamide (Temposil) or a placebo reduced their drinking over a four-month period. The fact that both the placebo and the drug produced the same results shows how patients' beliefs determine the effects of anti-alcohol drugs. However, many addiction specialists believe that anti-alcohol drugs can be helpful to some people, especially those who are motivated to abstain yet are fearful that they may be tempted to drink.

Some new "anti-craving" drugs have also shown promise in the treatment of alcoholism. One class of these drugs inhibits the uptake of the neuro-transmitter serotonin and, in controlled trials, males with mild to moderate drinking problems who were given these drugs have reported less interest drinking and less actual drinking than those given a placebo (Naranjo and Bremner, 1990). Some promising results using dopamine and the opiate antagonized naltrexone have also been reported (Volpicelli, Clay et al., 1994). The Canadian government approved the use of naltrexone for alcoholism treatment in early 1996.

General Counselling and Psychotherapy

We have already noted that counselling and therapy can mean many different things. The term "psychotherapy" generally refers to "talk" treatments by therapists with a psychoanalytic orientation. Psychotherapists generally seek to give patients insight into their problems and to help them resolve conflicts that originate in early childhood.

As with all "talk" therapies, the competence and personal qualities of therapists may have as much to do with the outcome than the use of specific techniques. One U.S. study showed that counsellors differed widely in competence and other qualities and that the more competent and empathetic counsellors had better results (Valle, 1981). Therapist qualities including experience also affect the outcomes of behavioral treatments (Spivak, Sanchez-Craig and Davila, 1974).

Alcohol Education

Knowledge may be power, but it does not necessarily translate into action. This is the conclusion from studies of the benefits of educating alcoholics about alcohol and its effects (Miller and Hester, 1986a). Education *per se* does not lead to reduced drinking or fewer drinking problems. This has been known for many years but many service providers continue to believe that education is important and provide it to their clients. Education for prevention has also had mixed reviews (see Chapters 10 and 12).

Confrontation

Images that come to mind at the mention of the word "confrontation" often involve highly emotional settings in which someone's faults are hammered home by one or more assailants. Confrontation in alcohol treatment is sometimes like this, especially in group therapy. Some therapists seem to believe that aggressive confrontations are sometimes needed to "break through the denial" of alcoholism.

There are no studies showing benefits from emotional confrontations in treatment. In fact one study has shown that alcoholics treated in groups with hostile, aggressive leaders have worse outcomes than others (Miller, Brown and Simpson, 1994). Confrontations involving the play back of video tapes made when subjects are drunk have also proven ineffective and lead to increased depression and lower self-esteem (Feinstein and Tamerin, 1972).

A less aggressive form of feedback about drinking norms and the ways in which drinking contributes to health problems does seem to influence people who are beginning to experience drinking problems. This was shown in a Swedish study in which healthy, middle-aged, male heavy drinkers, identified through liver tests, were given information about the health consequences of their drinking and about moderate drinking levels (Kristenson, Ohlin et al., 1983). Over a five-year follow-up period, many men reduced their drinking, especially those who were given a detailed physical examination, a comprehensive interview, specific advice on drinking and several follow-up appointments. This was one of the first studies to show that simple interventions can have a significant effect on some types of drinkers.

Family Therapy

Problem drinkers who have stable supportive relationships tend to do better after treatment than others (Moos, Finney and Cronkite, 1990) and many therapists try to strengthen family relationships to improve the chances of treatment success. One Canadian and a few American studies show benefits of family therapy but the results are not clear cut. In the Canadian study drinkers and their spouses were randomly assigned to receive eight sessions of family therapy or a single session of advice or information (Zweben, Pearlman and Li., 1988). Drinkers in both groups improved significantly over an 18-month period. There were also signs of improvement in marital adjustments in both groups. Unfortunately, this study did not involve a comparison group where spouses were not involved in treatment. Therefore, the real benefits of spousal involvement are difficult to identify.

Behavioral Treatment

We have noted a variety of fairly specific treatments that come under the broad heading of "behavioral." Many of these have been tested, alone or in combination, in carefully controlled trials. Many of these trials have produced positive results and, in general, behavioral treatments have received more empirical support than most other forms of treatment (Holder, Longabaugh et al., 1991). This is especially true for mild to moderately dependent drinkers, although successful behavioral treatments involving severely dependent people have also been reported.

For well-motivated people who do not have severe drinking problems, relatively brief behavioral interventions or even simple advice have proved quite effective (Edwards, Ordord et al., 1977; Bien Miller and Tonigan, 1993). Typically these involve 3 to 8 sessions with a therapist and the use of self-study materials. The basic steps involve: (i) identifying current drinking patterns and high-risk drinking situations; (ii) a period of abstinence to improve the drinker's health and thinking; (iii) setting long-term goals; and (iv) learning strategies to cope with high-risk situations and ways to use time formerly taken up with drinking. These approaches can also be helpful for people with more severe problems. However, more dependent people may require additional help to improve social, stress management and employment skills and other aspects of their lifestyles (Kadden, 1994).

Several researchers at the Addiction Research Foundation in Toronto have made important contributions to research on behavioral treatments. These include Martha Sanchez-Craig, who has done ground-breaking work on self-help and brief interventions and Helen Annis who has developed and evaluated widely respected relapse-prevention materials and strategies. Mark and Linda Sobell have also published many studies that show the benefits of behavioral interventions.

Self-Help Groups

Alcoholics Anonymous is the oldest and best known of the so-called 12-step groups that are currently in vogue for many types of problems. Less well-known, and far smaller, is Women for Sobriety. The U.S. groups Rational Recovery and Save Our Sobriety (SOS) have very few members in Canada.

Due to their informal nature, the effectiveness of self-help groups is impossible to determine and the question, "Do they work?" needs to be replaced by a series of more specific questions about the effects of particular types of self-help group experiences on particular individuals. Certainly more people with drinking problems go to self-help groups (especially to AA) than to specialized treatment services and those who continue in such groups presumably find them helpful. Stories of remarkable changes following involvement in AA are very common and, in follow-up studies, people who go to AA are less likely to continue drinking than others (Emrick, Tonigan et al., 1993). However many people who go to AA only attend a few meetings and do not follow the life-long 12-step program. They may nonetheless benefit from exposure to AA unless they come away convinced that they have no choices except to join AA or continue as problem drinkers.

Early studies suggested that people who become long-term members of AA tend to be "high-bottom" binge drinkers who had experienced many negative consequences as a direct result of drinking but who still had good social support and career options. These studies also suggested that AA appealed most to middle-aged males with a strong need to belong and with religious tendencies (Ogborne and Glaser, 1981). However, AA is now highly diversified and its appeal is obviously quite wide. Anyone who has a drinking problem should certainly go to a few AA meetings with an open mind to find out what it is all about.

There are no Canadian studies of the outcomes of people referred to AA by the courts, employers or professionals. Three American studies showed that people referred to AA by the courts or by employers were no more likely to reduce their drinking than others (Brandsma, Maultsby and Walsh, 1980; Ditman, Crawford et al., 1967; Walsh, Hingson et al., 1991). There are no outcome studies of other self-help groups in Canada.

Studies that suggest that increasing numbers of AA groups may contribute to decreases in liver-cirrhosis rates and to decreases in alcohol-related offences have been published using Canadian and American data (Smart, Mann and Anglin, 1989; Mann, Smart and Anglin,

1988). These are intriguing notions and suggest that AA may be having quite a powerful influence in some areas.

What are the Best Types of Treatment and Do They Work for Everyone?

Bearing in mind that many people overcome their problems without formal treatment, the treatment research literature clearly favors treatments that focus on drinking behaviors over those that seek to address underlying causes. However, these do not work for everyone and many people are clearly helped by other types of treatments. It is also possible that there are "different strokes for different folks" and this has given rise to the so-called "matching hypotheses," which states that treatment outcomes can be improved if people are matched to the treatment most suited to their needs and characteristics. This seemingly simple idea has attracted a good deal of interest among service providers. In Ontario, a system of assessment/referral services has been established, partly to match problem drinkers and drug users with appropriate resources in the community (Ogborne and Rush, 1990). Few studies have clearly demonstrated that client-treatment matching makes a large impact on treatment outcomes, although one frequently cited Canadian study did suggest that matching works.

This study was done at the Donwood Institute in Toronto by Dr. John McLachlan (McLachlan, 1972). It involved alcoholics who had been treated by therapists who were more or less "directive" in their treatment styles. The most directive therapists tended to have clear rules, to be very consistent and to run highly structured treatment sessions. Less directive therapists were more "laid back" and encouraged more self-expression and autonomy. By chance, some patients assigned to different types of therapists differed in what McLachlan called "conceptual level" (differences in personality and thinking styles resulting in differences in the need for structure and guidance).

McLachlan followed up patients treated by different therapists and found that those whose conceptual levels matched their therapist's styles did better than others. Thus, patients with a need for structure did best with directive therapists while those who were more self-directed fared better with a less structured approach. On the other hand, mismatched patients did less well with both types of therapists.

Dr. Helen Annis of the Addiction Research Foundation has also done research on matching and found that clients who drink in many different situations do equally well in structured relapse-prevention and counselling programs. Those whose drinking is more situation-specific do best with structured relapse prevention (Annis and Davis, 1988). Recent U.S. studies generally support the matching hypothesis and suggest that additional patient and treatment factors need to be considered (DHSS, 1993).

Does It Matter Where Treatment Takes Place?

Many people believe that it is important for problem drinkers to get away from their home environments to receive proper treatment. This is said to be for the benefit of the drinker, who can then relax and concentrate on recovery, and also for family members who are temporarily relieved of the strains of living with an active problem drinker. These benefits were heavily promoted by 19th-century sanatoriums and their rural locations were consistent with the "getting-away-from-it-all" philosophy.

Currently Canada has about 400 residential treatment centres that specialize in the treatment of addiction for the general population and 51 residential treatment centres for native people. These centres have generally not been evaluated using control or comparison group designs — that is to say designs that compare the outcomes of those treated in residential centres with those treated on an outpatient basis. American and English studies suggest that these comparisons would generally favor outpatient treatments, which tend to be equally effective and less expensive (Miller and Hester, 1986b). However, more research is needed because it is clear that residential centres have several functions including the provision of temporary relief to the families of chronic drinkers with a poor long-term prognosis. Long-term residential recovery homes also seem to be helpful for socially unstable people seeking to re-establish themselves.

Can Alcoholics Ever Become Social Drinkers?

We saw in Chapters 5 and 7 that many researchers question the usefulness of the term "alcoholic" and argue with the notion that there is a clear distinction between "alcoholics" and other heavy drinkers. We also learned that research shows that people move in and out of drinking problems and that the "progressive disease" model of alcoholism is an over simplification. Therefore, we can say with some confidence that many problem drinkers do drink socially from time to time. Many surveys, experimental studies and treatment-outcomes studies support this conclusion (Miller, 1983). However, this does not mean that social drinking is a realistic or desirable long-term goal for most people who are heavily dependent on alcohol. The risks of failure are high and there are often significant contra-indications.

Perhaps the best reason for severely dependent people to abstain completely is that they usually have a long history of failed attempts at social drinking. This is how some people define alcoholism. But even those who question the term "alcoholism" recognise that, for some people, social drinking inevitably leads to trouble. Exactly how and why may be an important scientific question, but for affected individuals and their families and friends, scientific explanations are less important than practical resolutions. Abstinence is a clear-cut goal and many options are available for those who commit to it. For long-term, alcohol-dependent drinkers and their families the alternative will usually involve constant vigilance and concern and styles of drinking that bring little pleasure. Of course, many long-term heavy drinkers should also try to remain abstinent because they have alcohol-related health problems. This is especially true for those with liver and cardiovascular problems.

Early-stage and less-dependent problem drinkers should also think seriously about abstinence. Although there may be health benefits of low to moderate consumption for some drinkers (see Chapter 7), the costs and risks of continued drinking for those with any reason for concern are considerable. In fact, when given a choice of abstinence or controlled drinking, some early stage drinkers chose abstinence (Sanchez-Craig, 1980). However, there is ample evidence that, if they wish, many of these people can learn to drink moderately and without problems.

One group that poses special challenges to treatment professionals is made up of severely dependent drinkers who do not accept the need to change their drinking habits. This includes people with serious health problems as well as those whose families despair over constant drunkenness and related problems. How can these people be helped?

There are no simple answers but one option is to find a common ground between committed drinkers and those who prefer them to change. This may involve, for example, an agreement not to drink and drive, not to drink every day and to reduce the number of drinks per day. Thus people who drink two bottles of wine a day may agree to set their limit at one. This is still heavy drinking, but a considerable improvement. They may also agree to take cabs after drinking rather than drive themselves. Other "risk-reduction" options that might be considered in individual cases include improvements in eating habits, attention to medical problems or drinking only at set times. If successful, these strategies may be a "foot in the door" to greater changes and to eventual abstinence. Certainly they are worth a try and preferable to turning people away from treatment if they do not immediately agree to demands of treatment professionals.

The Future of Treatment for Alcohol Problems in Canada

In the foreseeable future, it seems unlikely that Canadians will turn their backs on providing treatment services for people with alcohol problems. These services are widely recognized as important and effective, and many Canadians would like more such services (Rush and Tyas, in press). Also, provincial and other levels of governments have invested heavily in treatment facilities and relevant human resources.

However, it also seems unlikely that resources for treatment will be increased over the balance of the 1990s. Currently, resources for health and social services are stretched to the limit and severe cuts have been made in some sectors, particularly in the hospital sector. In this context, new addiction treatment services seem unlikely to be given a priority and there will be increased pressures on existing services to demonstrate their cost-effectiveness.

We are therefore likely to see an increased emphasis on accountability and program evaluation. This is already evident from the development of large-scale provincewide management information systems (Camp, Krakow et al., 1992) and increasing interests in evaluation among program managers. "How-to" books on evaluation for program managers are being published (e.g., Graham, Woo et al., 1994) and evaluation workshops are very popular. Large-scale economic evaluations of treatment are also being planned.

Recent evaluations of treatment have indicated that structured self-help guides and relatively brief interventions are sometimes effective especially for people who are not heavily dependent on alcohol (Bien, Miller and Tonigan, 1993). This has sparked an interest in promoting these interventions and in encouraging early-stage problem drinkers to use them. The expectation is that these approaches will reduce the short-term costs of service. They may also have long-term economic benefits if fewer drinkers progress to the point of where they need costly health and social services. This is also consistent with more general efforts to redirect some health care resources from treatment to prevention.

The search for new drugs for the treatment of alcoholism and other chemical addictions is quite active and the possibility of a drug "cure" excites some researchers, drinkers and members of the public.

However, others remain sceptical and recall that both morphine and LSD were once touted as cures for alcoholism. Drugs tend to have side effects that limit their value and make them unattractive to users. They must also be taken if they are to work and compliance with drug regimens is often limited, except for highly motivated people. It is also of note that sugar-coated pills can be very effective as "drugs" if they are believed to work.

Summary

Until the end of the Second World War, very few people with alcohol problems were provided with any type of treatment. The number of treatment programs increased after the war, especially during the late-1960s and the 1970s. Currently, a wide range of treatments are provided in a variety of specialized settings. General practitioners and other general health and social service providers are also involved in treating people with alcohol problems and especially those whose problems are less serious. However, most people with problems do not seek formal treatment, although many go to meetings of Alcoholics Anonymous. Research generally provides more support for behavioral treatment than for other types and there is no good evidence that longer-term or intensive treatments are necessarily more effective that brief, less-intensive treatments. However, treatment outcomes are heavily influenced by the characteristics of those being treated and the social environments to which they return. There is some evidence in favor of matching people with treatment according to their individual needs and circumstances. Promising research involving the use of anti-craving drugs has also been reported.

Chapter 12

What is Canadian Drinking, Eh? A Summary and Some Conclusions

When we planned the first edition of this book, we expected it to show how alcohol and its problems were wrapped up in both Canadian history and contemporary social life. We wanted to come to some conclusions about what was typical about Canadian drinking and our attempts to solve its problems. Indeed, it is still the case that a study of drinking touches on many of the recurrent themes in Canadian life. Such themes include problems with the North and its development, problems experienced by native people, and issues around the integration of Quebec into Canada's political and social life. Other typically Canadian themes are regionalism, recognition of our puritan past in a more liberal present, a strong dependence on government to solve problems and the Balkanization of political entities able to do so. In this summary chapter, we offer a few closing statements on what typifies both Canadian drinking and efforts to solve the resulting problems, and lastly, on what further work is most needed.

What Typifies Canadian Drinking and Its Problems?

Most Canadians Are Moderate Drinkers

Although most Canadians now drink alcoholic beverages, few do so very frequently. There are also clear signs that overall consumption and the percentage of heavy drinkers are both decreasing. However, most abstainers are relatively old and are not being replaced when they die. The next generation will therefore have a somewhat higher proportion of drinkers.

Only about 11 per cent of Canadians drink every day and only about 30 per cent drink once a week or more. Moreover, the average consumption is now only 10 drinks a week — about the middle of the range for western industrialized countries. Clearly, personal alcohol use is not generally a problem for most Canadians and there may be health and other benefits for some light drinkers. However, upwards of five per cent of Canadians are dependent on alcohol or have other significant alcohol-related problems. Also many Canadians have experienced occasional problems with alcohol and many have been affected by the drinking of other people.

Puritan Ideals and Liberal Views Conflict

Canada has had a long-standing tradition of puritan influence in both politics and social life. Many of its original settlers were the Methodists and Baptists who created the temperance movement. Canada is one of the few countries in the world that had peacetime Prohibition; but the way Prohibition was subverted by thirsty citizens exposed a real conflict with the puritan view.

Much the same conflict can be seen in modern society. Canada is one of the few countries to have monopoly systems for alcohol controls — not even the United States has it in most places. Although the Canadian system is being weakened by creeping privatization, most visitors from other countries claim that alcohol is less available here than elsewhere, and they are right. However, there is enormous social support for drinking. Alcohol is offered at most social and family gatherings, at many business functions, and at most government-sponsored official dinners. Even various provincial alcoholism commissions and foundations have permitted alcohol to be served at sponsored events. Despite the heavy controls on availability that were created after Prohibition, Canadians manage to drink a fairly large amount; as much as Americans, Australians and the British and far more than Swedes or Norwegians, who also had strong temperance movements. One might think that Canada's puritan past and low availability would insulate Canadians from this degree of consumption, but this is not so.

Drunkenness is Often Accepted

Some people have argued that in countries with conflicting attitudes to alcohol, there is more overt drunkenness and less daily drinking. Perhaps this conflict explains the large amount of drunkenness in Canada. Many recreational and social events involve heroic amounts of drinking. Weddings, birthdays, and New Year's Eve and Christmas parties often are occasions for heavy drinking. Drunkenness is frequently sought by teenage drinkers at weekend parties. Lately, complaints about drunkenness at baseball and football games have been so common that some people will not take children to them. Drunkenness is also very much a feature of festivities surrounding sports events such as the Grey Cup, the Stanley Cup and the World Series, where many of those involved are from the middle class.

There was a popular television skit in the 1980s involving "Bob and Doug Mackenzie" a pair of stuporous Canadian boys dressed in typical outdoor clothing. They sat, surrounded by beer, too drunk to even hold a conversation, but gamely trying. Bob and Doug later made a hit record called "The Beer Hunter," which describes a game where contestants open as much beer as possible in order to find one that does not effervesce. They must drink any beer they open, and become very drunk in doing so. A feature film has the same characters in a farcical plot centred on a brewery run by a mad scientist. In one scene a character saves the day by drinking a huge vat of beer. Doug and Bob seem to represent characters that many Canadians recognize or even sympathise with. Otherwise, it would be difficult to explain their popularity.

Native People Have Special Problems

Both Indians and Inuit have special problems with alcohol. Whether the main influences are social, historical or genetic is much debated. Most of the evidence, however, suggests that the main influences are environmental and social and hence open to change. Certainly, special efforts will be needed to ameliorate the problems native peoples have with alcohol and the broader social and economic issues that underlie them. As development of the North proceeds, these problems are sure to worsen. It is clear, though, that aboriginal peoples in the United States, Australia, New Zealand and New Guinea experience many of the same problems as Canadian native people. There may well be something about aboriginal status in the face of a dominant alien culture that predisposes to heavy drinking worldwide.

Regional Differences Predominate

Most social and cultural facts in Canadian life are strongly influenced by "regionalism." No matter what you study, people in the Maritimes are almost sure to do it differently from people in Yukon. Drinking is no exception to this rule. Of course, the original strength of the temperance movement was in the Maritimes and Ontario, and it never achieved much success in British Columbia, Quebec, or the North. In general, drinking is much heavier in the northern and western parts of Canada. People in the Maritimes drink somewhat less. Also, spirits are more popular in the West, and wine in Ontario. Beer is most popular in Quebec and Newfoundland. There are also differences in trends, as the North shows larger recent decreases in consumption and alcohol-related problems.

Quebec Stands Out as Different

Socially and culturally Quebec can be seen as very different from the remainder of Canada. Traditionally, these differences were very striking for all questions involving alcoholic beverages, and a number of differences remain even today. During the 1800s, French Quebec had its own small temperance movement and ignored the larger movements in English Canada. Yet its citizens consistently voted against Prohibition, and their vote led to its rejection on a national level by Laurier in 1898. During Prohibition in English Canada, people imported alcoholic beverages either legally or illegally from Quebec. Even today it is the place where thirsty Ontarians go to stock up during beer strikes. Quebec has had a more liberal approach to alcohol controls than other provinces and represents something of a Canadian bastion of freedom in such matters. For example, only Quebec allows the sale of beer and wine in grocery stores. Other provinces, as they did during Prohibition, covet this greater freedom. In general, however, the amount of drinking in Quebec is not very different from that in most other provinces, except the Maritimes.

Frontier Drinking Still Exists

Few countries in the world still have isolated frontiers, but Canada is certainly one of them. Historical examples of frontier drinking among railway men, lumbermen, miners, and prospectors are easy to find, and many of these examples have modern counterparts. Vast areas of land in the North are still frontiers in that the population is sparse, there are no cities, and social organization is often weak. These areas often depend upon resource-based economic development, and hence attract many hard-drinking young single men. Because these men have high incomes, leisure time, and few recreational facilities, drinking assumes a great importance in the social life of frontier areas. Frontier drinking tends to be heavy, oriented to getting drunk, and has a desperate escapist quality for isolated males.

As the North is developed, drinking problems could become worse, and special steps would be needed to combat them. Of course, part of the problem alcohol causes in Indian and Inuit communities results from a frontier drinking style. Frequently, native groups live in areas undergoing development and are exposed to drinking styles brought by workers from the South. Development projects also mean that some native people get jobs at high wages and hence have more money for alcohol.

What Typifies Canadian Efforts to Deal with Drinking Problems?

It is relatively easy to characterize Canadian drinking. So many empirical studies have been done over a long time that most aspects of drinking are well understood. However, it is more difficult to describe what is typical in the way drinking problems are handled. Some features do stand out — especially our traditional dependence on treatment approaches, our desire to have governments do the work, and our abhorrence of anything suggesting Prohibition.

We Have Traditionally Depended on Treatment

The predominant mode of response to drinking problems has been to provide treatment for alcoholics. Until recently, all provinces (except Quebec) had alcoholism commissions or foundations whose main job was to provide treatment for alcoholics. Most did little else, although the Foundations in Ontario and Alberta did a considerable amount of prevention work, and the Addiction Research Foundation of Ontario has always done a large amount of research. However, prevention programs such as alcohol education in schools, mass media efforts, and community-based programs have been developed in other parts of the country over the last decade. Many employers have developed programs to identify and assist employees with drinking problems and efforts to prevent drinking and driving are evident in most communities. General "health promotion" campaigns are also much in evidence across the country and these typically seek to dissuade people from excessive drinking. These are all positive developments because it is unlikely that drinking problems will be solved if we only aim to treat people with drinking problems and do nothing to prevent these problems from developing in the first place. The National Drug Strategy and several provincial strategy documents demonstrate some movement away from exclusive dependence on treatment and more emphasis on prevention.

We Expect Governments to Do It

Generally, Canadians like, or at least tolerate, a large degree of government involvement in their lives. They expect governments to solve social problems rather than see them as the responsibility of individuals, families, or private enterprise. For example, all provinces have prepaid government-run health services that have largely prevented the development of private health services for alcoholics and others.

Governments are the sole distributors and sellers of alcoholic beverages but, at the same time, they are expected to solve the resultant problems. This conflict has often been seen as the reason why so little is done. Real prevention may result in fewer sales and less tax revenue to governments constantly searching for new tax bases. Given the problems with unemployment in the farm and tourism sectors, governments may be reluctant to take any measures that may make things worse.

Prevention and Treatment Efforts are not Well Co-ordinated Nationally

In Canada, both health services and the distribution of alcoholic beverages are largely provincial matters. Consequently, there are 12 provincial and territorial systems of health service and 12 alcohol distribution systems. The distribution systems seem to rarely interact, but often compete, and there appears to be no overall Canadian system of alcohol controls. Nor is there any Canadian approach to the prevention of alcohol problems or their treatment. Each province and territory largely goes its own way, and there is lack of co-ordination in all efforts at prevention and treatment.

The federal government, through Health Canada, does promote some mass media education and some community alcohol prevention programs. Also, the federally funded Canadian Centre on Substance Abuse has become an important forum for the exchange of ideas and information. However, there is no strong federal presence setting standards or guidelines for prevention or treatment of alcohol problems.

Neo-temperance Approaches Are Rejected

The disappearance of Prohibition left Canadians with a strong distaste for strict controls on their access to alcohol. Anything that smacks of Prohibition is likely to be rejected out of hand. Indeed, even small changes in regulations are criticized in the media as "a return to Prohibition." Most people want to have good access to alcoholic beverages at convenient times and at reasonable prices. Unfortunately, the bad memories of Prohibition have sometimes made it difficult for governments to be open to the best ideas for prevention. For example, much research indicates that heavy drinking can best be reduced by lowering per capita consumption through increases in price and reductions in availability. This approach has been labelled "neo-prohibitionist" or "neo-temperance" by its opponents. Of course, there is some truth in the implied charge; the general approach is not very different from that argued by temperance groups. However, temperance groups argued for alcohol restraints largely for moral reasons, whereas the modern scientific approach seeks the prevention of health-related problems. Considerable changes in public attitudes about alcohol would be necessary before there was much acceptance of any genuinely "neo-temperance" approach. For example, 34 per cent of people in Ontario want alcohol taxes reduced and only 18 per cent want taxes to be increased for public health reasons. Others (52 per cent) want taxes to remain the same. (Ferris et al., 1994).

Alcohol Problems and Alcohol Availability Are not Widely Seen as Related

Most Canadians abhor heavy drinking, drunkenness, impaired driving, and all of the other negative consequences of drinking. However, there seems to be no generally held view about how those problems arise. Many people seem to see alcoholics as totally different from heavy social drinkers in terms of their sensitivity to alcohol. Most people seem to think that the level of drinking in the country is not related to the rate of alcohol problems. For example, there are frequent references, both by politicians and the press, to the merits of French drinking styles. Also, the general availability of alcohol in the public mind is not linked to the frequency of problems. For example, strikes affecting the beer or liquor industries tend to get front-page coverage and are presented as provincial disaster. Booze is rushed from foreign countries and other provinces. Although strikes could be seen as having some public health benefits, this is rarely discussed.

There are, however, signs that attitudes are changing. For example, public and professional opinion lead the 1985-1990 Ontario Liberal government to back away from an election promise to allow beer and wine to be sold in liquor stores. The Ontario Ministry of Health has also recently developed a "population-based'" substance-abuse prevention and treatment strategy. This is based on the explicit assumption that "normal drinking" and "problem drinking" are related, and one objective of the strategy is to reduce per capita alcohol consumption.

What Are the Greatest Needs for the Future?

Most social problems, including those related to alcohol, need so much study and programming that a quick summary is impossible.

However, some of the most important needs for addressing alcohol problems stand out clearly. They are:

- continued research on problems among high-risk groups such as young people, native people, and Northerners;

- continued monitoring of trends in overall drinking, styles of drinking, and attitudes toward our own and other people's drinking;

- efforts to reduce per capita alcohol consumption and the resulting problems. These efforts should include increasing the price of alcoholic beverages and creating mass media programs to tell people why their own and the public health is well served by these changes;

- continued provision of low-cost, effective treatments for alcoholics and other problem drinkers, together with research on how to improve such treatments. In many areas an expansion of treatment facilities will be needed. Generally, the expansion of treatment services is followed by decreases in alcohol problems;

- the development of good alcohol education programs for schools, high-risk youth, drinking drivers and, especially, heavy drinkers;

- massive efforts (using techniques perhaps not yet available) to educate the public, especially heavy drinkers, journalists, and politicians, about the close link between population drinking levels, availability, and alcohol problems;

- greater efforts to identify and intervene early in alcohol-related problems, especially in employee assistance programs and physician-based programs; and

- more emphasis on separating drinking from driving especially for young people. Measures should be aimed at delaying young people's drinking, licensing them only to drive at certain times, and reducing the maximum legal blood alcohol level for young drivers.

References

Addiction Research Foundation (1984). *Statistics on Alcohol and Drug Use in Canada and Other Countries*. Toronto: Addiction Research Foundation.

Addiction Research Foundation (1995). *Response to Consultation Paper on Options to Reform the Criminal Code of Canada*. Toronto: Addiction Research Foundation.

Adlaf, E. M., Ivis, F. J. and Smart, R. G. (1994). *Alcohol and Other Drug Use among Ontario Adults in 1994 and Changes since 1977*. Toronto: Addiction Research Foundation.

Adlaf, E. M., Ivis, F. J., Smart R. G. and Walsh, G. W. (1995). *Ontario Student Drug Use Survey, 1977-1995*. Toronto: Addiction Research Foundation.

Adlaf, E. M. and Smart, R. G. (1985). Drug use and religious affiliation, feelings and behavior. *British Journal of Addiction* 80: 163-171.

Adlaf, E. M. and Smart, R. G. (1995). Alcohol use, drug use and well being in older adults. *International Journal of the Addictions* 30 (14/15): 1985-2016.

Adlaf, E. M., Smart, R. G. and Canale, M. D. (1991). *Drug Use Among Ontario Adults 1977-1991*. Toronto: Addiction Research Foundation.

Adlaf, E. M., Smart, R. G. and Tan, S. H. (1989). Ethnicity and drug use: A critical look. *International Journal of the Addictions* 24: 1-18.

Adlaf, E. M., Smart, R. G., and Walsh, G. W. (1992). Substance use and work disabilities among a general population. *American Journal of Drug and Alcohol Abuse* 18: 371-387.

Adlaf, E. M., Smart, R. G. and Walsh, G. W. (1993). *Ontario Student Drug Use Survey*. Toronto: Addiction Research Foundation.

Adrian, M. and Williams, R. T. (1989). *Do Indians Increase Alcohol Consumption in Ontario?* Toronto: Addiction Research Foundation, Internal Document #102.

Affleck. (1877). *Matchless Gems of Stirring Song for Public and Social Gatherings, Schools, Conventions and Camp Meetings*. Toronto: A Millerd.

Albert, W. G. and Simpson, R. I. (1985). Evaluation of an educational program for the prevention of impaired driving among Grade 11 students. *Journal of Drug Education* 15: 57-71.

Albert, W. G., Simpson, R. I. and Eaglesham, J. (1983). Evaluation of a drinking and parenting educational program in six Ontario communities. *Journal of Drug Education* 13: 327-335.

Allen, H. T. (1981). *Forty Years' Journey: The Temperance Movement in British Columbia to 1900*. Victoria: B.C. Historical Society.

Allen, K. (1994). Development of an instrument to identify barriers to treatment for addicted women, from their perspective. *International Journal of the Addictions* 29: 429-444.

American Psychiatric Association (1994). *Diagnostic and Statistical Manual* 4th ed. Washington, DC: American Psychiatric Association.

Anderson, P. (1980-81). Alcoholism and the spinal cord disabled: A model program. *Alcohol Health and Research World* 8: 37-41.

Anderson, P. (1994). Overview: Public health, health promotion and addictive substances. *Addiction* 89: 1523-1527.

Anglin, L., ed. (1995). *The Ontario Experience of Alcohol and Tobacco: New Focus on Accessibility, Violence and Mandatory Treatment*. Toronto: Addiction Research Foundation, Internal Document No. 122.

Anglin, L., Mann, R. E. and Smart, R. G. (1995). Changes in cancer mobility rates and per capita Alcohol consumption in Ontario, 1963-1987. *International Journal of the Addictions* 30: 489-495.

Annis, H. M. and Davis, C. S. (1988). Relapse prevention. In *Comprehensive Handbook of Alcoholism Treatment Approaches*, ed. Hester, R. K., Miller, W. R. and Lettieri, D. New York: Pergamon Press.

Anton, R. F. (1994). Medications for treating alcoholism. *Alcohol Health and Research World* 18: 265-271.

Archibald, H. D. (1990). *The Addiction Research Foundation: A Voyage of Discovery*. Toronto: Addiction Research Foundation.

Ashley, M. J. (1992). Alcohol-related birth defects. In *Aboriginal Substance Use: Research Issues*, ed. McKenzie, D. Toronto: Canadian Centre for Substance Abuse.

Ashley, M. J., Ferrence, R., Room, R., Rankin, J. and Single, E. (1994). Moderate drinking and health: A report of an international symposium. *Canadian Medical Association Journal* 151: 1-20.

Ashley, M. J., le Riche, W., Olin, J., Hatcher, J., Kornaczewski, A., Schmidt, W. and Rankin, J. (1978). 'Mixed' (drug abusing) and 'pure' alcoholics: A socio-medical comparison. *British Journal of Addiction* 73: 19-34.

Ashley, M. J., Olin, J. S., le Riche, W., Harding, A., Kornaczewski, A., Schmidt, W. and Rankin, J. G. (1977). Morbidity in alcoholics: Evidence for accelerated development of physical disease in women. *Archives of Internal Medicine* 137: 883-887.

Ashley, M. J., Olin, J. S., le Riche, W. H., Kornaczewski, A., Schmidt, W., Corey, P. N. and Rankin, J. G. (1981). The physical disease characteristics of inpatient alcoholics. *Journal of Studies on Alcohol* 42: 1-14.

Assembly of First Nations (1988). *Current Indian Health Conditions: A Statistical Perspective*. Unpublished Report.

Aylward, S. (1994). Women's participation and recovery in AA. Paper presented at a conference on Addiction and Mutual Help Movements in a Comparative Perspective. Toronto, September 1994.

Barbary, H. (Undated). Alcohol, drugs and violence. In *Violence in Contemporary Canadian Society*, ed. MacLatchie, J. M. Toronto: The John Howard Society of Canada.

Barnes, J. G. and Bourgeois, J. C. (1977). *Factors Which Influence Per Capita Consumption of Beverage Alcohol*. Ottawa: Non-medical Use of Drugs Directorate.

Barron, F. L. (1976). *The Genesis of Temperance in Ontario, 1828-1850*. Ph.D. thesis, University of Guelph.

Beck, A. T., Weissman, A. and Kovacs, M. (1976). Alcoholism, hopelessness and suicidal behavior. *Journal of Studies on Alcohol* 37: 66-67.

Beck, K. H. and Summons, T. G (1987). The social context of drinking among high school drinking drivers. *American Journal of Drug and Alcohol Abuse* 13: 181-198.

Bell, R. G. and Soloman, S. (1989). *A Special Calling: My Life in Addiction Treatment and Care*. Toronto: Stoddart.

Berger, D. E. and Snortum, J. R. (1985). Alcoholic beverage preferences of drinking-driving violators. *Journal of Studies on Alcohol* 46: 232-239.

Berger, T. R. (1977). *Northern Frontier, Northern Homeland: The Report of the Mackenzie Valley Pipeline Inquiry*, Vol. 1. Ottawa: Ministry of Supply and Services, Canada.

Bien, T. H., Miller, W. R. and Tonigan, J. S. (1993). Brief interventions for alcohol problems: a review. *Addiction* 88(3): 315-335.

Blane, H. T. and Hewitt, L. E. (1977). *Mass Media Public Education and Alcohol: A State of the Art Review*. Pittsburgh: University of Pittsburgh.

Bohman, M. (1978). Some genetic aspects of alcoholism and criminality: A population of adoptees. *Archives of General Psychiatry* 35: 269-276.

Bohman,M., Sivardsson, S. and Cloninger, C. R. (1981). Maternal inheritance of alcohol abuse: Cross fostering analysis of adopted women. *Archives of General Psychiatry* 38: 965-969.

Boland, B. and Roizen, R. (1973). Sales slips and survey responses: New data on the reliability of survey consumption measures. *Drinking and Drug Practices Surveys* 8: 5-10.

Bondy, S. (1994). *Attitudes and Experiences with Treatment of Alcohol and Tobacco Problems: A Report of the Ontario Alcohol and Other Drug Opinion Survey, 1993*. Toronto: Addiction Research Foundation.

Brandsma, J. M., Maultsby, M. C. and Walsh, R. J. (1980). *Outpatient Treatment of Alcoholism: A Review and Comparative Study*. University Park Press: Baltimore MD.

Brazeau, R. and Burr, N. (1992). *International Survey: Alcoholic Beverage Taxation and Control Policies*. 8th ed. Ottawa: Brewers Association Of Canada.

Briggs, S. R. (1986). *Trinity of Evil*, Toronto.

Brown, M., Devan, M. F. and Wallace, P. (1982). *International Survey: Alcoholic Beverage Taxation and Control Policies*. 5th ed. Ottawa: Brewers Association of Canada.

Brown, R. C. and Cook, R. (1974). *Canada: 1896-1912. A Nation Transformed*. Toronto: McClelland & Stewart.

Bruun, K., Edwards, G., Lumio, M., Makela, K., Pan, L., Popham, R., Room, R., Schmidt, W., Skog, O. J., Sulkunen, P. and Osterberg, E. (1975). *Alcohol Control Policies in Public Health Perspective*. Toronto: The Finnish Foundation for Alcohol Studies.

Camp, J. M., Krakow, M., McCarty, D. and Argeriou, M. (1992). Substance abuse treatment management information systems: Balancing federal, state and service providers needs. *The Journal of Mental Health Administration* 19: 5-19.

Carlen, P. L., Wilkinson, A. D., Wortzman, G., Holgate, R., Cordingley, J., Lee, M. A., Huszar, L., Moddel, G., Singh, R., Kiraly, L. and Rankin, J. G. (1981). Cerebral atrophy and functional deficits in alcoholics without clinically apparent liver disease. *Neurology* 31: 377-385.

Carpenter, E. S. (1959). Alcohol in the Iroquois dream quest. *American Journal of Psychiatry* 116: 148-151.

Caverson, R. (1990). A programmer's perspective on implementing a drinking-driving countermeasure at the community level. In *Research Action and the Community: Experiences in the Prevention of Alcohol and Other Drug Problems*, ed. Giesbrecht, N., Conley, P., Denniston, R., Gliksman, L., Holder, H., Pederson, A., Room, R. and Shain, M. Rockville, MD: Office for Substance Abuse Prevention.

Charlebois, C. A. and Poel, D. H. (1982). *Drug Awareness Week in Nova Scotia-1981 — Community Context and Campaign Impact*. Halifax: Dalhousie University.

Cherpitel, C. J. (1994). Alcohol and causalities: A comparison of emergency room and coroner data. *Alcohol and Alcoholism* 29: 211-218.

Cherrington, E. H. (1928). *Standard Encyclopedia of the Alcohol Problem*. Westerville, OH: American Issue Publishing Co.

Cheung, Y. W. (1990). Ethnicity and alcohol/drug use revisited: a framework for future research. *International Journal of the Addictions* 25: 581-603.

Churley, M. (1994). LCBO fights smuggled and illegally manufactured alcohol. (Press release) Toronto: Ministry of Consumer and Commercial Relations.

Cloninger, C. R., Bohman, M. and Sigvardsson, S. (1981). Inheritance of alcohol abuse: Cross fostering analysis of adopted men. *Archives of General Psychiatry* 38: 861-868.

Coade, E. (1974). Suicide and suicide attempts in Montgomery County Maryland. Montgomery, MD: Montgomery County Office of Drug Control.

College of Family Physicians of Canada (1994). *Alcohol Risk Assessment and Intervention* (ARAI). Ottawa: College of Family Physicians of Canada.

Cross, M. S. (1960). *The Lumber Community of Upper Canada*. Ontario Historical Society.

Dailey, R. C. (1968). The role of alcohol among North American Indian tribes as reported in the Jesuit Relations. *Anthropoligica* 10: 45-59.

Dean, J. C., Fox, A. M. and Jensen, B. A. (1985). Drug and alcohol use by disabled and nondisabled persons: A comparative study. *The International Journal of Addictions* 20: 629-641.

Delin, C. R. and Lee, T. H. (1992). Drinking and the brain: current evidence. *Alcohol and Alcoholism* 27: 117-126.

deLint, J. and Levinson, T. (1975). Mortality among patients treated for alcoholism: A five-year follow-up. *Canadian Medical Association Journal* 13: 385-387.

deLint, J. and Schmidt, W. (1968). The distribution of alcohol consumption in Ontario. *Quarterly Journal of Studies on Alcohol* 29: 968-973.

Department of Indian and Northern Affairs (1981). *Indian Conditions: A Survey*. Ottawa: Ministry of Indian and Northern Affairs (QS-5141-0000-EE-A3).

Dewar, R. and Sommer, R. (1964). Alcohol consumption in Prairieville. *Quarterly Journal of Studies on Alcohol* 25: 300-313.

Ditman, K. S., Crawford, G. G., Forgy, E. W., Moskowitz, H. et al. (1967). A controlled experiment on the use of court probation for drunk arrests. *American Journal of Psychiatry* 124: 160-163.

Douglas, R. R. (1990). Formulating alcohol policies for community recreation facilities: Tactics and problems. In *Research Action and the Community: Experiences in the Prevention of Alcohol and Other Drug Problems,* ed. Giesbrecht, N., Conley, P., Denniston, R., Gliksman, L., Holder, H., Pederson, A., Room, R. and Shain, M. Rockville: Office for Substance Abuse Prevention.

Drummond, A., Cave, T. and Healy, D (1987). The risk of accident involvement by time of week: An assessment of the effect of zero BAC legislation and the potential of driving curfews. In *Young Drivers Impaired by Alcohol and Other Drugs*, ed. Benjamin, T. London: Royal Society of Medicine Services.

Dunning, R. S. (1993). Trade issues and alcohol monopolies: An international perspective. In *International Symposium on Alcohol Monopolies and Social Health Issues,* ed. Ferris, J. et al. Toronto: Addiction Research Foundation.

Duster, T. (1983). Commentary. In *Alcohol and Disinhibition: Nature and Meaning of the Link,* NIAAA Research Monograph No. 12, ed. Room, R. and Collins, G. Washington, DC: USGPO, DHHS Publication No. (ADM) 83-1246, 326-330.

Edmonton Sun (1994). Safeway Rethinks Booze Sales, 2 June.

Edwards, G., Anderson, P., Babor, T. F., Casswell, S., Ferrence, R., Giesbrecht, N., Godfrey, C., Holder, H. D., Lemmens, P., Mäkelä, K., Midanik, L. T., Norström, T., Österberg, E., Romelsjö, A., Room, R., Simpura, J. and Skog, O. J. (1994). *Alcohol Policy and the Public Good.* Oxford: Oxford University Press.

Edwards, G., Kyle, E., Nicholls, P. and Taylor, C. (1978) Alcoholism and correlates of mortality: Implications for epidemiology. *Journal of Studies on Alcohol* 39: 1607-1617.

Edwards, G., Ordord, J., Egert, S., Guthrie, S., Hawker, A., Hensman, C., Mitcheson, M., Oppenheimer, E. and Taylor, C. (1977). Alcoholism: A controlled trial of "treatment" and "advice." *Journal of Studies on Alcohol* 38: 1004-1031.

Ehline, D. and Tinghe, P. O. (1977). Alcoholism: Early identification and intervention in the social service agency. *Child Welfare* 56: 384-592.

Elgee, W. H. (1964). *The Social Teaching of the Canadian Churches*. Toronto: Ryerson Press.

Eliany, M., Giesbrecht, N., Nelson, M., Wellman, B. and Wortley, S. eds. (1990). *National Alcohol and Drugs Survey (1989): Highlights Report*. Ottawa: Health and Welfare Canada.

Eliany, M., Giesbrecht, N., Nelson, M., Wellman, B. and Wortley, S. (1992). *Alcohol and Other Drug Use by Canadians: A National Alcohol and Other Drugs Survey (1989) Technical Report*. Prepared for Health and Welfare Canada.

Eliany, M. and Rush, B. (1992). *How Effective are Alcohol and Other Drug Prevention and Treatment Programs?: A Review of Evaluation Studies*. A Canada's Drug Strategy Baseline Report #3. Ottawa: Health and Welfare Canada.

Eliany, M., Giesbrecht, N., Nelson, M., Wellman, B. and Wortley, S. eds. (1990). *National Alcohol and Drugs Survey (1989): Highlights Report*. Ottawa: Health and Welfare Canada.

Elliott, D. and Cosper, R. (1982). The time-budget study of tavern-going: A validation. *Journal of Studies on Alcohol* 43: 397-403.

Ellis, K. and Rush, B. R. (1993). *Alcohol and Other Drug Services in Ontario: Results of a Provincial Survey, 1992*. Toronto: Addiction Research Foundation.

Emrick, C. D., Tonigan, J. S., Montgomery, H. and Little, L. (1993). Alcoholics Anonymous: What is currently known. In *Research on Alcoholics Anonymous: Opportunities and Alternatives*, ed. McCrady, B. S., Miller, W. R. New Brunswick, NJ: Rutgers Centre for Alcohol Studies.

Eye Disease Case-control Study Group. (1993). Risk factors for branch retinal vein occlusion. *American Journal of Ophthalmology* 116: 286-296.

Farmer, P. J. (1975). The Edmonton study: A pilot project to demonstrate the effectiveness of a public information campaign on the subject of drinking and driving. In *Alcohol, Drugs and Traffic Safety*, ed. Israelstam, S. and Lambert, S. Toronto: Addiction Research Foundation, 831-843.

Feinstein, C. and Tamerin, J. S. (1972). Induced intoxication and videotape feedback in alcoholism treatment. *Quarterly Journal of Studies on Alcohol* 33: 85-90.

Ferris, J., Room, R. and Giesbrecht, N. (1993). Public health interests in trade agreements on alcoholic beverages in North America. *Alcohol, Health and Research World* 17: 235-241.

Ferris, J. and Templeton, L. (1993). *Ontario Alcohol and Other Drug Survey Report 1992*. Toronto: Addiction Research Foundation.

Ferris, J., Templeton, L. and Wong, S. (1994). *Alcohol, Tobacco and Marijuana: Use, Norms, Problems and Policy Attitudes among Ontario Adults*. Toronto: Addiction Research Foundation, Internal Document No. 118.

Fiddler, S. (1985). *Suicides, Violent and Accidental Deaths Among Treaty Indians in Saskatchewan*. Federation of Saskatchewan Indian Nations.

Fingarette, H. (1988). *Heavy Drinking: The Myth of Alcoholism as a Disease*. Berkley: University of California Press.

Fisher, A. D. (1987). Alcoholism and race: The misapplication of both concepts to North American Indians. *Canadian Review of Sociology and Anthropology* 24: 81-97.

Forbes, E. R. (1971). Prohibition and the social gospel in Nova Scotia. *Acadiensis* 1: 11-36.

Frank, D. H. (1983). "If you drink, don't drive" motto now applies to hangovers as well. *Journal of the American Medical Association* 250: 1657-1658.

Frankel, B. E., Ferrence, R., Johnson, F E. and Whitehead, P. C. (1976). Drinking and self injury: Toward untangling the dynamics. *British Journal of Addiction* 71: 299-306.

Fried, P. A. and Watkinson, B. (1988). Twelve and 24-month neurobehavioural follow-up of children prenatally exposed to marijuana, cigarettes and alcohol. *Neurotoxicol Teratol* 10: 305-313.

Friendenreich, C. M., Howe, G. R., Miller, A. B. and Jain, M. G. (1993). A cohort study of alcohol consumption and risk of breast cancer. *American Journal of Epidemiology* 137: 512-520.

Gallup (1943-1995). *The Gallup Poll*. Toronto: Gallup Canada, Inc.

Garceau, S. (1985). *A Summary Report on Tobacco, Alcohol and Marijuana Use Among Young People in Canada*. Ottawa: Health Promotion Directorate, Health Canada.

Garland, M. A., and Talman, J. J. (1931). Pioneer drinking habits and the rise of the temperance agitation in Upper Canada prior to 1840. Toronto: Ontario Historical Society, Papers and Records XXI.

Gavin, M., Bohm, P., Carpen, R., DeVilliar, M. et al. (1987). Final report of the treatment services coordinating and advisory committee detox working group. Toronto: Addiction Research Foundation, Internal document No. 87.

Gemini North Ltd. (1974). Social and Economic Impact of the Proposed Arctic Gas Pipeline in Northern Canada. Prepared for Canadian Arctic Gas Pipeline Ltd.

Gerson, L. W. and Preston, D. A. (1979). Alcohol consumption and the incidence of violent crime. *Journal of Studies on Alcohol* 40: 307-312.

Gervais, G. H. (1980). *The Rumrunners: A Prohibition Scrapbook*. Thornhill, ON: Firefly Books.

Gfellner, B. M. and Hundleby, J. D. (1995). Patterns of drug use among native and white adolescents: 1990-1993. *Canadian Journal of Public Health* 86: 95-97.

Giesbrecht, N., Brown, J., deLint, J. and Lambert, S. (1977). *Alcohol Problems in Northern Ontario: A Preliminary Report*. Toronto: Addiction Research Foundation, Substudy #872.

Giesbrecht, N., Conley, P., Denniston, R. W., Gliksman, L., Holder, H., Pederson, A., Room, R. and Shain, M., eds. (1990). *Research Action and the Community: Experiences in the Prevention of Alcohol and Other Drug Problems*. Rockville: Office for Substance Abuse Prevention.

Giesbrecht, N. and Macdonald, S. (1981). A ban on fortified wine in northwestern Ontario and its impact on the consumption level and drinking patterns. *British Journal of Addiction* 76: 281-288.

Giesbrecht, N., Pranovi, P. and Wood, L. (1990). Impediments to changing local drinking practices: Lessons from a prevention project. In *Research Action and the Community: Experiences in the Prevention of Alcohol and Other Drug Problems*, ed. Giesbrecht, N., Conley, P., Denniston, R., Gliksman, L., Holder, H., Pederson, A., Room, R. and Shain, M. Rockville, MD: Office for Substance Abuse Prevention.

Gliksman, L., Adlaf, E. M., DeWit, D., Giesbrecht, N. and Newton-Taylor, B. (1994). Trends in drug use and problems among Ontario's university students. Paper presented at the 20th Annual Alcohol Epidemiology Symposium of the Kettil Bruun Society for Social and Epidemiological Research on Alcohol. Ruschlikon, Switzerland, June 6-10.

Gliksman, L., Douglas, R. R. and Smythe, C. (1983). The impact of a high school drug education program using a live theatrical performance: A comparative study. *Journal of Drug Education* 13: 229-247.

Gliksman, L., Engs, R. and Smythe, C. (1989). *The Drinking, Drug Use, and Lifestyle Patterns of Ontario's University Students*. Toronto: Addiction Research Foundation.

Gliksman, L., Hart, D., Simpson, R. and Siess, T. (1990). *Progress on Campus: Evaluation of the Campus Alcohol Policies and Education (CAPE) Program*. Toronto: Addiction Research Foundation.

Gliksman, L. and Single, E. (1988). A field evaluation of a server intervention program: Accommodating reality. Paper presented at the Canadian Evaluation Society Meeting, Montreal, 1988.

Globe and Mail (1994). Mohawk smuggling continues to boom. June 29.

Goodstadt, M. S. and Sheppard, M. A. (1983). Three approaches to alcohol education. *Journal of Studies on Alcohol* 44: 362-380.

Goodstadt, M. S., Sheppard, M. A. and Chan, G. C. (1982). An evaluation of two school-based alcohol education programs. *Journal of Studies on Alcohol* 43: 352-369.

Goodstadt, M. S., Smart, R. G. and Gillies, M. (1978). Public attitudes toward increasing the price of alcoholic beverages. *Journal of Studies on Alcohol* 39: 1630-1632.

Goodwin, D. W., Schulsinger, F., Hermansen, L., Guze, S. B. and Winokur, G. (1973). Alcohol problems in adoptees raised apart from biological parents. *Archives of General Psychiatry* 31: 164-169.

Goodwin, D. W., Schulsinger, F., Moller, N., Mednick, S. and Guze, S. B. (1977). Psychopathology in adopted and non-adopted daughters of alcoholics. *Archives of General Psychiatry* 34: 1005-1009.

Gottfried, E. B., Korsten, M. A. and Liebes, C. S. (1976). Gastritis and duodenitis induced by alcohol: an endoscopic and hutologic assessment (abstract). *Gasterenterology* 70(A32), 890.

Gowen, H (1899). *Church Work in British Columbia*. London: Longmans Green.

Graham, K. (1984). Determinants of heavy drinking and drinking problems: The contribution of the bar-environment. Paper presented at the symposium on Public Drinking and Public Policy, Banff, Alberta, April 1984.

Graham, K. (1986). Identifying and measuring alcohol abuse among the elderly: Serious problems with existing instrumentation. *Journal of Studies on Alcohol* 47: 322-326.

Graham, K., Carver, V. and Brett, P. (1992). Alcohol and drug use by older women. Paper presented at the 18th Annual Alcohol Epidemiology Symposium of the Kettil Bruun Society. Toronto, June 1-5.

Graham, K., Woo, G. et al. (1994). *The Evaluation Casebook*. Toronto: Addiction Research Foundation.

Graham-Clay, S. (1983). Fetal alcohol syndrome: A review of the current human research. *Canada's Mental Health* 31: 2-5.

Gray, J. H. (1972). *Booze: The Impact of Whisky on the Prairie West*. Toronto: Macmillan.

Gray, J. H. (1982). *Bachanalia Revisited: Western Canada's Boozy Skid to Social Disaster*. Saskatoon, SK: Western Producer Prairie Books.

Guillet, E. C. (1954). *Pioneer Inns and Taverns*. Toronto: published by the author.

Guillet, E. C. (1979). *Pioneer Days in Upper Canada*. 1933. Reprint, Toronto: University of Toronto Press.

Gusfield, J. R. (1981). *The Culture of Public Problems: Drinking-Driving and the Symbolic Order*. Chicago & London: University of Chicago Press.

Hallowell, G. A. (1972). *Prohibition in Ontario, 1919-1923*. Vol. 2. Toronto: Ontario Historical Society.

Hawthorn, H. B., Belshaw, C. S. and Jamieson, S. M. (1957). The Indians of British Columbia and alcohol. *Alcoholism Review* 2: 10-14.

Health and Welfare Canada (1984). *Alcohol in Canada: A National Perspective*. Ottawa: Health and Welfare Canada.

Heath, A. C., Jardine, R. and Martin, N. G. (1989). Interactive effects of genotype and social environment on alcohol consumption in female twins. *Journal of Studies on Alcohol*, 50: 38-48.

Heath, D. W. (1982). Alcohol and drug use among North American Indians: A cross-cultural survey of patterns and problems. In *Research Advances in Alcohol and Drug Problems*. Vol. 7. Ed. Smart, R. G., Glaser, F., Israel, Y., Popham, R. E. and Schmidt, W. New York: Plenum Press.

Hepner, R., Kirshbaum, H. and Landes, D. (1980). Counselling substance abusers with additional disabilities: The centre for independent living. *Alcohol Health and Research World* 5: 11-15.

Hickey, W. (1834). *Hints on Emigration to Upper Canada*. London.

Hiebert, A. J. (1969). *Prohibition in British Columbia*. Master's thesis, Simon Fraser University.

Hingson, R., Heeren, T. and Morelock, S. (1986). Preliminary effects of Maine's 1982 0.02 law to reduce teenage driving after drinking. In *Young Drivers Impaired by Alcohol and Other Drugs*, ed. Benjamin, T. London: Royal Society of Medicine Services.

Hobart, C. W. (1978). Economic development, liquor consumption and offence rates in the Northwest Territories. *Canadian Journal of Criminology* 20: 257-279.

Holder, H. D. (1993). The store monopoly as a public policy approach to consumption and alcohol problems: A review of research evidence. In *International Symposium on Alcohol Monopolies and Social Health Issues*, ed. Ferris, J. et al. Toronto: Addiction Research Foundation, 117-140.

Holder, H., Longabaugh, R., Miller, W. and Rubonis, A. V. (1991). The cost effectiveness of treatment for alcoholism: A first approximation. *Journal of Studies on Alcohol* 52: 517-525.

Holder, H. and Wagenaar, A. C. (1994). Mandated server training and reduced alcohol-involved crashes: a time series analysis of the Oregon experience. *Accident Analysis and Prevention* 26: 89-97.

Howard-Pitney, B., Johnson, M. D., Altman, D. G., Hopkins, R. and Hammond, N. (1991). Responsible alcohol service: A study of server manager and environmental impact. *American Journal of Public Health* 81: 197-199.

Howay, F. W. (1942). The introduction of intoxicating liquors to the Indians of the Northwest Coast. *B.C. Historical Quarterly* 4: 157-169.

Howe, J. (1973). *Western and Eastern Rambles: Travel Sketches of Nova Scotia*. Toronto: University of Toronto Press.

Howison, J. (1821). *Sketches of Upper Canada*. Edinburgh: Oliver and Boyd.

Hubert, K. (1990). *IMPACT: Measuring Success*. Edmonton: Alberta Alcohol and Drug Commission.

Hyland, J. and Scott, S. (1969). *Alcohol Consumption Tables*. Toronto: Addiction Research Foundation.

International Agency for Research on Cancer (1988). *IARC Monographs on the Evaluation of Carcinogenic Risks to Human: Alcohol drinking*, Vol. 44. Geneva: World Health Organization.

James, I. P. (1966). Blood alcohol levels following successful suicide. *Journal of Studies on Alcohol* 27: 29-33.

Jameson, A. (1943). *Winter Studies and Summer Rambles in Canada*. Toronto: Thomas Nelson.

Jellinek, E. M. (1960). *The Disease Concept of Alcoholism*. New Brunswick, NJ: Hillhouse Press.

Jones, B. M. and Parson, D. A. (1977). Impaired abstracting ability in alcoholics. *Archives of General Psychiatry* 24: 71-75.

Johnson, J. A. and Oksanen, E. H. (1974). Socio-economic determinants of consumption of alcoholic beverages. *Applied Economics* 6: 293-301.

Johnson, S. D., Gibson, L. and Linden, R. (1978). Alcohol and rape in Winnipeg, 1966-1975. *Journal of Studies on Alcohol* 39 (1): 1887-1894.

Johnston, L. D., O'Malley, P. M. and Bachman, J. G. (1992). *Drug Use, Drinking and Smoking: National Survey Results from High School, College and Young Adult Populations, 1975-1991*. Washington, DC: National Institute on Drug Abuse.

Jung, J. (1994). *Under the Influence: Alcohol and Human Behavior*. Belmont, CA: Books/Cole.

Kadden, R. M. (1994). Cognitive-behavioral approaches to alcoholism treatment. *Alcohol Health and Research World* 18: 279-286.

Kaprio, J., Koskenvuo, M., Langinvaino, H., Ramonov, K., Sarna, S. and Rose, R. J. (1987). Genetic influences on use and abuse of alcohol: A study of 5,638 adult Finnish twin brothers. *Alcoholism: Clinical and Experimental Research* 11: 349-356.

Kendler, K. S., Heath, A. C., Neale, M. C., Kessler, R. C. and Eaves, L. J. (1992). A population-based twin study of alcoholism in women. *Journal of the American Medical Association* 268: 1877-1882.

Kilty, K. M. (1990). Drinking styles of adolescents and young adults. *Journal of Studies on Alcohol* 51: 556-564.

King, W. L. M. (1942). *Temperance and a Total War Effort*. Ottawa: King's Printer.

Kinney, J., Leaton, G. (1995). *Loosening the Grip: A Handbook of Alcohol Information*, 5th ed. St. Louis: Mosby.

Kissin, B. (1974). Interactions of ethyl alcohol and other drugs. In *The Biology of Alcoholism*, Vol. 3., ed. Kissin, B. and Begleiter, H. New York: Plenum Press.

Kleisinger, G. J. (1984). *The Safe Grad Program*. Regina.

Kohn, P. M. and Smart, R. G. (1984). The impact of television advertising on alcohol consumption: an experiment. *Journal of Studies on Alcohol* 45: 295-301.

Kohn, P. M. and Smart, R. G. (1987). Wine, women, suspiciousness and advertising. *Journal of Studies on Alcohol* 48: 161-166.

Kohn, P. M., Smart, R. G. and Ogborne, A. C. (1984). Effects of two kinds of advertising on subsequent consumption. *Journal of Advertising* 13: 34-40.

Korsten, M. A. and Wilson, J. S. (1993). Alcohol and the pancreas: Clinical aspects and mechanisms of injury. *Alcohol Health and Research World* 15: 292-298.

Kristenson, H., Ohlin, H., Hulten-Nosslin, B., Trell, E. and Hood, B. (1983). Identification and intervention of heavy drinking in middle-aged men: Results and follow-up of 24-60 months of long-term study with randomized controls. *Alcoholism: Clinical and Experimental Research* 7: 203-209.

Lalinec-Michaud, M., Subak, M. E., Ghadrian, A. M. and Kovess, V. (1991). Substance misuse among native and rural high school students in Quebec. *International Journal of the Addictions* 26: 1003-1012.

Langevin, R. (1982). The role of alcohol, drugs, suicide attempts and situational strains in homicide committed by offenders seen for psychiatric assessment: A controlled study. *Acta Psychiatrical Scandinavia Supplement* 66: 229-242.

Leacock, S. (1919). The warning of Prohibition in America. *National Review London* 73: 680-689.

Ledermann, S. (1956). *Alcool, alcoolisme, alcoolisation: Données scientifiques de caractère physiologique, économique et social.* (Institut National d'Etudes Démographiques, Travaux et Documents, Cahier No. 29). Paris: Presses Universitaires de France.

Legge, C. and Sherlock, L. (1990-91). Perception of alcohol use and misuse in three ethnic communities: Implications for prevention programming. *International Journal of the Addictions* 25: 629-653.

Lemmens, P. H. (1994). The alcohol content of self report and standard drinks. *Addictions* 89: 413-426.

Lex, B. W. (1991). Some gender differences in alcohol and polysubstance users. *Health Psychology* 10: 121-132.

Liban, C. B. and Smart, R. G. (1982). Drinking and drug use among Ontario Indian students. *Drug and Alcohol Dependence* 9: 161-171.

Lin, T. and Lin, D. T. (1982). Alcoholism among the Chinese: Further observations of a low-risk population. *Culture Medicine & Psychiatry* 6 (2). 109-116.

Listiak, A. (1974). "Legitimate deviance" and social class: Bar behavior during Grey Cup Week. *Sociological Focus* 7: 13-44.

Lithman, Y. G. (1979). Feeling good and getting smashed: On the symbolism of alcohol and drunkenness among Canadian Indians. *Ethnos* 44: 119-133.

Lockwood, E. C. (1993). Temperance in Canada as an ethnic subterfuge. *In Drink in Canada: Historical Essays,* ed. Warsh, C. K. Montreal: McGill-Queen's University Press.

MacAndrew, C. and Edgerton, R. B. (1969). *Drunken Comportment: A Social Explanation*. Chicago: Aldine Publishing.

Macdonald, S. and Wells, S. (1993). *Employee Assistance, Health Promotion and Drug Testing Programs at Ontario Work Sites*. Toronto: Addiction Research Foundation.

Macdonald, S. (1995). The role of drugs in workplace injuries: Is drug testing appropriate? *Journal of Drug Issues* 25(4): 703-723.

MacKinnon, A A. and Newfeld, A. H. (1973). *Project Mental Health: A Study of Opinion North of 60 Degrees*. Canadian Mental Health Association, Northwest Territories Division.

Maclean's (1994). Booze and the badge: a report suggests the RCMP has alcohol problems. *Maclean's* 107(13), March 28: 52-55.

MacMahon, S. (1987). Alcohol consumption and hypertension. *Hypertension* 9(2). 111-121.

Mäkelä, K., Room, R., Single, E., Sulkunen, P. and Walsh, B. with 13 others (1981). *Alcohol, Society, and the State: 1. A Comparative Study of Alcohol Control*. Toronto: Addiction Research Foundation, 1981.

Makowsky, C. R. and Whitehead, P. (1991). Advertising and alcohol sales: a legal impact study. *Journal of Studies on Alcohol* 52: 555-567.

Mann, R. E., Vingilis, E. R., Leigh, G., Anglin, L. and Blefgen, H. (1986). School-based programs for the prevention of drinking and driving: issues and results. *Accident Analysis and Prevention* 18: 325-337.

Mann, R. E., Smart, R. G. and Anglin, L. (1988). Are increases in liver cirrhosis rates a result of increased treatment for alcoholism? *British Journal of Addiction* 83: 683-688.

Maracle, J. (1993). *Crazywater: Native Voices on Addiction and Recovery*. Toronto: Viking.

May, P. A. (1989). Alcohol abuse and alcoholism among American Indians: An Overview. In *Alcoholism in Minority Populations*, ed. Watts, T. O. and Wright, R. Springfield, IL: Charles C. Thomas, 95-119.

Mayfield, D. E. and Montgomery, D. (1972). Alcoholism, alcohol intoxication and suicide attempts. *Archives of General Psychiatry* 27: 349-353.

McClelland, D. C., Davis, W. N., Kalin, R. and Warner, E. (1972). The drinking man. New York: Free Press.

McClung, Nellie (1935). *Clearing in the West: My Own Story*. Toronto: Thomas Allen.

McCord, W. and McCord, J. (1960). *Origins of Alcoholism*. Stanford CA: Stanford University Press.

McGregor, J. (1833). *British America*, London.

McKenzie, D., ed. (1992). *Aboriginal Substance Use: Research Issues*. Toronto: Canadian Centre on Substance Abuse and National Native Alcohol and Drug Abuse Program,.

McKnight, A. J. (1987). *Development and Field Test of a Responsible Alcohol Service Program, Vol. I, Research Findings*. Ottawa: National Highway Traffic Safety Administration, Report No. DOT HS 807 221.

McLachlan, J. F. C. (1972). Benefit from group therapy as a function of patient-therapist match on conceptual level. *Psychotherapy: Theory Research and Practice* 9: 317-323.

McLean, R. F. (1969). *A "Most Effectual Remedy": Temperance and Prohibition in Alberta, 1875-1915*. Master's thesis, University of Calgary.

McLellan, A. T., Grossman, D. S., Blaine, J. D. and Haverkos, H. W. (1995). Dr. A. Thomas McLellan and colleagues, response regarding evidence for effectiveness of acupuncture. *Journal of Substance Abuse Treatment* 12(3): 141-148.

Menninger, K. A. (1938). *Man Against Himself*. New York: Harcourt Brace.

Merritt, J. P. (1875). *Biography of the Hon. W. H. Merritt*. St. Catharines.

Merskey, H., Brant, C. C. and Malla, A. (1988). Symptom patterns of alcoholism in northern Ontario population, *Canadian Journal of Psychiatry* 33: 46-50.

Meyers, A. R., Branch, L. G., Lederman, R. I. (1988). Alcohol, tobacco and cannabis use by independently living adults with major disabling conditions. *International Journal of the Addictions* 23: 671-685.

Miller, W. R. (1983). Controlled drinking: A history and a critical review. *Journal of Studies on Alcohol* 44, 68-83.

Miller, W. R., Brown, J. M., Simpson, T. L. et al. (1994). What works? A methodological analysis of the alcohol treatment outcome literature. In *Handbook of Alcoholism Treatment Approaches*, 2nd edition, ed. Hester, R. K. and Miller, W. R. New York: Plenum Press.

Miller, W. R. and Hester, R. K. (1986a). The effectiveness of alcoholism treatment: what research reveals. In *Treating Addictive Behaviors: Processes of Change*, ed. Miller, W. R and Heather, N. New York: Plenum Press.

Miller, W. R. and Hester, R. K. (1986b). Inpatient alcoholism treatment: who benefits. *American Psychology* 41: 794-805.

Moodie, S. (1959). *Life in the Clearings*, ed. McDougall, R. L. 1853. Reprint, Toronto: University of Toronto.

Moodie, S. (1980). *Roughing it in the Bush*. 1852. Reprint, Toronto: Coles Publishing.

Moos, R. H., Finney, J. W. and Cronkite, R. (1990). *Alcoholism Treatment: Context, Process and Outcome*. New York: Oxford University Press.

Murphy, G. E., Armstrong, J. W., Hermele, S. L., Fischer, J. R. and Clendenin, W. W. (1979). Suicide and alcoholism. *Archives of General Psychiatry* 36: 65-69.

Murray, R. (1839). *A Course of Lectures on Absolute Abstinence*. Toronto.

Naranjo, C. A. and Bremner, K. E. (1992). Evaluation of the effects of seratonin uptake inhibitors in alcoholics: A review. In *Novel Pharmacological Interventions For Alcoholism*, ed. Naranjo, C. A. and Sellers, E. M. New York: Springes-Veslag.

National Association of Friendship Centres (1985). *Urban Research Project, Phases I & II; Alcohol and Solvent Abuse*. Unpublished Report.

National Native Advisory Council on Alcohol and Drug Abuse (1983). *Proceedings of the National Symposium on Fetal Alcohol Syndrome, March 14-15*, 1983.

Native Women's Association of Canada (1981). *Report for the National Native Alcohol Abuse Program*. Unpublished report.

Nelson, J. P. (1990). State monopolies and alcoholic beverage consumption. *Journal of Regulatory Economics* 2: 83-98.

Nestle, M. (1994). *Le paradoxe francaise*. Report presented at an International Symposium on Moderate Drinking and Health. Toronto, 29 April, 1993.

Newman, Peter C. (1978). *Bronfman Dynasty: The Rothschilds of the New World*. Toronto: McClelland and Stewart.

Noel, I. (1993). Dry patriotisms: The Cliniquy crusade. In *Drunk in Canada: Historical Essays*, ed. Warsh, C. K. Montreal and Kingston: McGill-Queen's University Press.

Noel, I. (1994). Temperance campaigning and alcohol consumption: a case study from pre-Confederation Canada. *Contemporary Drug Problems* 21: 401-426.

O'Dowd, B. F., Rothhammes, F. and Israel, Y. (1990). Genotyping of mitochondrial aldehyde dehydrogenase locus of Native American Indians. *Alcoholism: Clinical and Experiment Research* 14: 531-533.

Ogborne, A. C., Adrian, M., Newton-Taylor, B. and Williams, R. (1991). Long-term trends in male drunkenness arrests in Metro Toronto: Effects of social-setting dexotication centers. *American Journal of Drug and Alcohol Abuse* 17(2). 187-198.

Ogborne, A. C. and Glaser, F. B. (1981). Characteristics of affiliates of Alcoholics Anonymous: A review of the literature. *Journal of Studies on Alcohol* 42: 661-675.

Ogborne, A. C. and Rush, B. R. (1990). Specialized addictions assessment/referral services in Ontario: A review of their characteristics and other in the addiction treatment system. *British Journal of Addiction* 85: 197-204.

Ogborne, A. C. and Smart, R. G. (1980). Will restrictions on alcohol advertising reduce alcohol consumption? *British Journal of Addiction* 75: 293-296.

Ogborne, A. C., Sobell, M. S. and Sobell, L. C. (1995). The significance of environmental factors for the design and evaluation of alcohol treatment programs. In *Determinants of Substance Abuse: Biological, Psychological and Environmental Factors*, ed. Galizio, M. and Maisto, S. A. New York: Plenum Press, 363-383.

Oliver, E. H. (1923). *The Liquor Traffic in the Prairie Provinces.* Toronto: Presbyterian Board of Home Missions and Social Service.

Ormsby, M. A. (1958). *British Columbia: A History*. Toronto: Macmillan.

Orrego, H., Blake, J. K., Blendis, L. M., Compton, K. V. and Israel, Y. (1987). Long term treatment of alcoholic liver disease with propylthiouracil. *New England Journal of Medicine* 317: 1421-1427.

Ossenberg, R. J. (1969). Social class and bar behavior during an urban festival. *Human Organizations* 28: 29-34.

Parker, D. A. and Harford, T. C. (1992). The epidemiology of alcohol consumption and dependence across occupations in the United States. *Alcohol Health & Research World* 16 (2). 97-105.

Parker, D. A. and Harman, M. S. (1978). The distribution or consumption model of prevention of alcohol problems. *Journal of Studies on Alcohol* 39: 377-399.

Peachey, J. E., Annis, H. A., Bornstein, E. R., Maglana, S. M. and Sykora, K. (1989). Calcium carbimide in alcoholism treatment. *British Journal of Addiction*, 84: 1359-1366.

Pentz, M. A., Dwyer, J. H., Mackinnon, D. P., Flay, B. R., Hansen, W. B. and Wang, E. Y. (1989). A multicommunity trial for primary abuse of adolescent drug abuse. *Journal of the American Medical Association* 261: 3259-3266.

Pernanen, K. (1974). Validity of survey data on alcohol use. In *Research Advances in Alcohol and Drug Problems*, Vol. 1., ed. Gibbins, R. J., Israel, Y., Kalant, H., Popham, R. E., Schmidt, W. and Smart, R. G. New York: Wiley.

Pernanen, K. (1976). Alcohol and crimes of violence. *In The Biology of Alcoholism*, Vol. 4., ed. Kissin, B. and Begleiter, H. New York: Plenum Press.

Pernanen, K. (1991). *Alcohol in Human Violence.* New York: Guilford Press.

Perrine, M. W. (1975). The Vermont driver profile: A psychometric approach to early identification of potential high-risk drinking drivers. In *Alcohol Drugs and Traffic Safety*, ed. Israelstam, S. and Lambert, S. Toronto: Addiction Research Foundation, 199-223

Pickens, R. W., Svikis, D. S., McGue, M., Lykken, D. T., Heston, L. L. and Clayton, P. J. (1991). Heterogeneity in the inheritance of alcoholism: A study of male and female twins. *Archives of General Psychiatry* 48: 19-28.

Pickering, J. (1831). *Inquiries of an Emigrant, Being the Narrative of an English Farmer, from the Year 1824 to 1830*. London: New Edition.

Pinno, E. (1971). *Temperance and Prohibition in Saskatchewan*. Master's thesis, University of Saskatchewan, Regina.

Potter, J. P. and McMichael, A. J. (1984). Alcohol, beer and lung cancer — a meaningful relationship. *International Journal of Epidemiology* 13: 240-280.

Popham, R. E. (1962). The urban tavern: Some preliminary observations. *Addictions* 9: 16-28.

Popham, R. E. (1976). Working papers on the history of the taverns, 2 — Legislative history of the Ontario taverns: 1774-1974. Toronto: Addiction Research Foundation, Substudy No. 809.

Popham, R. E. (1982). Working papers on the tavern: Notes on the contemporary tavern. Toronto: Addiction Research Foundation.

Popham, R. E. and Schmidt, W. (1981). Words and deeds: The validity of self-report data on alcohol consumption. *Journal of Studies on Alcohol* 42: 355-358.

Popham, R., Schmidt, W. and deLint, J. (1978). Government control measures to prevent hazardous drinking. In *Drinking: Alcohol in American Society: Issues and Current Research*, ed. Ewing, J. A. and Rouse, D. A. Chicago: Nelson-Hall.

Popham, R. E., Schmidt, W. and Israel, Y. (1983). Variations in mortality from ischemic heart disease in relation to alcohol and milk consumption. *Medical Hypotheses* 12: 321-329.

Powell, K. C. (1988). *Drinking and Alcohol in Colonial Australia 1788-1901 for the Eastern Colonies*. Canberra: Australian Government Publishing Service.

Preusser, D., Williams, A. and Zador, P. (1984). The effects of curfew laws on motor-vehicle crashes. *Law and Policy* 6: 125-128.

Produktschap voor Gedistilleerde Dranken [PGD] (1993). *World Drink Trends*. Henly on Thames, UK: NTC Publications Ltd.

Proudfoot, W. (Undated). *Diary of William Proudfoot*. Unpublished.

Quarterly Journal of Inebriety (1892). *Quarterly Journal of Inebriety*, 96.

Radford, J. L., King, A. J. C. and Warren, W. K. (1989). *Street Youth and AIDS*. Ottawa: Health and Welfare Canada.

Rankin, J. (1994). Underlying biologic mechanisms. Report at an International Symposium on Moderate Drinking and Health. Toronto, April 29, 1993.

Ratcliffe, W. D. and Caverhill, K. M. (1978). Types of patrons in Edmonton beverage rooms. Unpublished manuscript.

Ratcliffe, W. D., Nutter, R. W., Hewitt, D., Gruber, G., Flanders, P. L. and Caverhill, K. M. (1982). Drinking in taverns: A 15-year comparison. *International Journal of the Addictions* 17: 869-877.

Ratcliffe, W. D., Nutter, R. W., Hewitt, D., Flanders, P. L., Caverhill, K. M. and Gruber, G. P. (1980). Drinking behavior in lounges and taverns. Unpublished manuscript. Edmonton: Alberta Alcoholism and Drug Abuse Commission.

Ray, A. J. (1974). *Indians in the Fur Trade: Their Role as Trappers, Hunters and Middlemen in the Lands Southwest of Hudson's Bay, 1660-1870*. Toronto: University of Toronto Press.

Reid, A. E. (1978). Alcohol and drug treatment in Canada: Programs, problems and prospects. Paper presented at a conference of the Canadian Addiction Foundation, Toronto.

Reid, J. A. (1976). *A Study of Limited Availability of Alcohol*. Regina: Alcoholism Commission of Saskatchewan.

Rex, D. K., Bosron, W. F., Smialek, J. E. and Li, T.-K. (1985). Alcohol and aldehyde dehydrogenace Isoenzymes in North American Indians. *Alcohol: Clinical and Experimental Research* 9: 147-152.

Rich, E. E. (1967). *The Fur Trade and the North-West to 1857*. Toronto: McClelland and Stewart.

Riddell, W. R. (1920). The first Canadian war-time prohibition measure. *Canadian Historical Review* 1: 187-190.

Rochemontaix (1919). Prohibition in Canada. Quoted in Spence, R. E. *Prohibition in Canada*. Toronto: Ontario Branch of the Dominion Alliance.

Room, R. (1990). Measuring alcohol consumption in the United States: methods and rationales. In *Research Advances in Alcohol and Drug Problems*, ed. Kozalowski, L. T. et al. New York: Plenum Press.

Room, R. (1993). The evolution of alcohol monopolies and their relevance for public health. *Contemporary Drug Problems* 20: 169-187.

Room, R. (1996). Patterns of family responses to alcohol and tobacco problems. *Drug and Alcohol Review* 15: 171-181.

Rose, A. W. (1849). *A Pioneer of the Wilderness: The Emigrant Churchman in Canada*. London.

Ross, A. (1972). *The Red River Settlement: Its Rise, Progress and Present State*. Edmonton: Hurtig.

Ross, H. E., Glaser, F. B. and Germanson, T. (1988). The prevalence of psychiatric disorders in patients with alcohol and other drug problems. *Archives of General Psychiatry* 45: 1023-1031.

Ross, H. E., Glaser, F. B. and Stiasny, S. (1988). Sex differences in the prevalence of psychiatric disorders in patients with alcohol and drug problems. *British Journal of Addiction* 83: 1179-1192.

Rush, B. R. and Ellis, K. (1993). *Alcohol and Other Drug Services in Ontario: Results of a Provincial Study, 1992*. Toronto: Addiction Research Foundation.

Rush, B. R., Gliksman, L. and Brook, R. (1983). *Determinants of Alcohol-related Damage in Large Populations: An Investigation Using Analysis of Linear Structural Relations (LISREL)*. Toronto: Addiction Research Foundation.

Rush, B. R. and Ogborne, A. C. (1992). Alcoholism treatment in Canada: History, current status and emerging issues. In *Cure, Care or Control: Alcoholism Treatment in Sixteen Countries*, ed. Klingemann, H., Takala, J. P. and Hunt, G. New York: State University Press.

Rush, B. R., Steinberg, M. and Brook, R. (1986). The relationships among alcohol availability, alcohol consumption and alcohol-related damage in the province of Ontario and the State of Michigan 1955-1982. *Advances in Alcohol and Substance Abuse* 4: 33-45.

Rush, B. R. and Tyas, S. (In press). Utilization and perceived value of alcohol and drug treatment services in Canada.

Ryan, C. and Butters, N. (1983). Cognitive deficits in alcoholics. In *The Biology of Alcoholism,* Vol. 7. Ed. Kissin, B. and Begleites, H. New York: Plenum Press, 485-538

Saltz, R. F. (1987). The role of bars and restaurants in preventing alcohol-impaired driving: An evaluation of server intervention. *Evaluation and the Health Professions* 10: 5-27.

Sanchez-Craig, M. (1980). Random assignment to abstinence or controlled drinking in a cognitive-behavioral program: Short-term effects on drinking behavior. *Addictive Behaviors* 5: 35-39.

Saskatchewan Alcohol & Drug Abuse Commission (1988). *Suicide in Saskatchewan*. Regina: SADAC.

Schlegel, R. P., Mariske, S. R. and Page, A. (1984). A guided decision-making program for elementary school students: a field experiment in alcohol education. In *Prevention of Alcohol Abuse,* ed. Miller, P. M. and Nirenberg, T. D. New York: Plenum Press.

Schmidt, W. (1977). The epidemiology of cirrhosis of the liver: A statistical analysis of mortality data with special reference to Canada. In *Alcohol and the Liver,* ed. Fisher, M. M. and Rankin, J. G. New York: Plenum Press.

Schmidt, W. and deLint, J. (1972). Causes of death of alcoholics. *Quarterly Journal of Studies on Alcohol* 33: 171-185.

Schmidt, W. and Popham, R. E. (1976). Impressions of Jewish alcoholics. *Journal of Studies on Alcohol* 37 (7). 931-939.

Scott, K. A. (1992). Substance use among indigenous Canadians. In *Aboriginal Substance Use: Research Issues,* ed. McKenzie, D. Toronto: Canadian Centre for Substance Abuse.

Seeley, J. R. (1960). Death by liver cirrhosis and the price of beverage alcohol. *Canadian Medical Association Journal* 83: 1361-1366.

Sellers, E. M. and Naranjo, C. A. (1983). Diazepam loading: Simplified treatment of alcohol withdrawal. *Clinical Pharmacology and Therapeutics* 34(6). 822-826.

Serles, J. S. (1988). The role of genetics in the pathogenesis of alcoholism. *Journal of Abnormal Psychology* 97: 153-167.

Shain, M. and Groenveld, J. (1980). *Employee Assistance Programs*. New York: Lexington Books.

Siggner, A. J. (1988). *Special Study on Youth*. Ottawa: Health Promotion Directorate, Health Canada.

Single, E. W. (1979). The "substitution hypothesis" reconsidered: A research note concerning the Ontario beer strikes in 1958 and 1968. *Journal of Studies on Alcohol* 40, 485-491.

Single, E. W., Brewster, J., MacNeil, P. and Hatcher, J. (1994). *Alcohol and Drug Use: Results from the General Social Survey, 1993*. Ottawa: Health Canada.

Single, E. W. and Giesbrecht, N. (1979). The 16 per cent solution and other mysteries concerning the accuracy of alcohol consumption estimates based on sales data. *British Journal of Addiction* 74: 165-173.

Single, E. W., Robson, L., Xie, X. and Rehm, J. (1996). *The Costs of Substance Abuse in Canada*. Ottawa: Canadian Centre on Substance Abuse.

Single, E. W., Williams, R. and McKenzie, D. (1994). *Canadian Profile: Alcohol, Tobacco and Other Drugs*. Toronto: Canadian Centre on Substance Abuse.

Single, E. W. and Wortley, S. (1993). Drinking in various settings as it relates to demographic variables and level of consumption: Findings from a national survey in Canada. *Journal of Studies on Alcohol* 54: 590-599.

Single, E. W. and Wortley, S. (1994). A comparison of alternative measures of alcohol consumption in the Canadian National Survey of alcohol and drug use. *Addiction* 89, 395-399.

Skinner, H. A. (1979). Profiles of treatment-seeking populations. Toronto: Addiction Research Foundation, Substudy #1054.

Skinner, H. A. and Allen, B. (1982). Alcohol-dependence syndrome: measurement and validation. *Journal of Abnormal Psychology* 91: 199-209.

Skinner, H. A., Holt, S. and Israel, Y. (1981). Early identification of alcohol abuse. 1. Critical issues and psychosocial indicators for a composite index. *Canadian Medical Association Journal* 124: 1141-1152.

Skinner, H. A., Reed, P. L. and Jackson, D. N. (1976). Toward the objective diagnosis of psychopathology: Generalizability of modal personality profiles. *Journal of Consulting and Clinical Psychology* 44: 111-117.

Smart, R. G. (1972). Observations of tavern patrons before and after the compulsory breath test law in Canada. *Journal of Studies on Alcohol* 33: 1122-1128.

Smart, R. G. (1974). Comparisons of purchasing in self-service and clerk service liquor stores. *Quarterly Journal of Studies on Alcohol* 35: 1397-1401.

Smart, R. G. (1977). Effects of two liquor store strikes on drunkenness, impaired driving and traffic accidents. *Journal of Studies on Alcohol* 38: 1785-1789.

Smart, R. G. (1979a). A note on the effects of changes in alcohol control policies in the Canadian North. *Journal of Studies on Alcohol* 40: 908-913.

Smart, R. G. (1979b). Female and male alcoholics in treatment: characteristics at intake and recovery rates. *British Journal of Addiction* 74, 275-281.

Smart, R. G. (1986). The impact on consumption of selling wine in grocery stores. *Alcohol and Alcoholism* 21, 233-236.

Smart, R. G. (1987). Socio-Economic, Lifestyle, and Availability Factors in the Stabilization of Alcohol Consumption in Canada. *Canadian Journal of Public Health*, 78, pp. 176-180.

Smart, R. G. (1988). Does alcohol advertising affect overall consumption? A review of empirical studies. *Journal of Studies on Alcohol* 49, 314-323.

Smart, R. G. (1989). Is the Postwar Drinking Binge Ending? Cross-National Trends in Per Capita Alcohol Consumption. *British Journal of Addiction*, 84, pp. 743-748.

Smart, R. G. (1990). Health warning labels for alcoholic beverages in Canada. *Canadian Journal of Public Health* 90, 280-284.

Smart, R. G. (1992). Observations of tavern patrons before and after the compulsory breath test law in Canada. *Quarterly Journal of Studies on Alcohol* 33, 1122-1128.

Smart, R. G. (1993). *The Smart Report: Substance Abuse and Canadian Youth*. Addiction Research Foundation, Toronto.

Smart, R. G. and Adlaf, E. M. (1986). Banning happy hours: The impact on drinking and impaired-driving dangers in Ontario, Canada. *Journal of Studies on Alcohol* 47: 256-258.

Smart, R. G. and Adlaf, E. M. (1987). Age-of-majority cards and drinking among young people. *Journal of Alcohol and Drug Education* 32: 60-64.

Smart, R. G., Adlaf, E. M., Porterfield, K. M. and Canale, M. (1990). *Drugs, Youth and the Street*. Toronto: Addiction Research Foundation.

Smart, R. G., Adlaf, E. M. and Walsh, G. W. (1991). *The Ontario Student Drug Use Survey: Trends Between 1977-1991*. Toronto: Addiction Research Foundation.

Smart, R. G., Adlaf, E. M. and Walsh, G. W. (1993). The relationship between declines in drinking and alcohol problems among Ontario students: 1979-1991. *Journal of Studies on Alcohol* 55: 338-341.

Smart, R. G., Adlaf, E. M., Walsh, G. W. and Zdanowicz, Y. M. (1992). *Drifting and Doing: Changes in Drug Use Among Toronto Street Youth, 1990-1992*. Toronto: Addiction Research Foundation.

Smart, R. G. and Cutler, R. E. (1976). Alcohol advertising ban in British Columbia: Problems and effects on beverage consumption. *British Journal of Addiction* 71: 13-21.

Smart, R. G. and Docherty, D. (1976). Effects of introduction of on-premise drinking on alcohol-related accidents and impaired driving. *Journal of Studies on Alcohol*, 37: 683-686.

Smart, R. G. and Goodstadt, M. S. (1977). Effects of reducing the legal alcohol purchasing age on drinking and drinking problems. *Journal of Studies on Alcohol* 38: 1313-1323.

Smart, R. G., Goodstadt, M., Adlaf, E. M., Sheppard, M. S. and Chan, G. C. (1984). Preliminary report of alcohol and drug use among Ontario students in 1983 and trends since 1977. Unpublished manuscript. Toronto: Addiction Research Foundation.

Smart, R. G. and Gray, G. (1979). Parental and peer influences as correlates of problem drinking among high school students. *International Journal of the Addictions* 14: 905-917.

Smart, R. G., Gray, G. and Bennett, C. (1978). Predictors of drinking and signs of heavy drinking among high school students. *International Journal of the Addictions* 13: 1079-1094.

Smart, R. G. and Liban, C. B. (1981). The need for attitude changes concerning drinking and drinking problems. *Journal of Alcohol and Drug Education* 27: 47-61.

Smart, R. G. and Liban, C. B. (1982). Alcohol consumption as estimated by the informant method: A household survey and sales data. *Journal of Studies on Alcohol* 43: 1020-1027.

Smart, R. G. and Mann, R. E. (1990). Changes in suicide rates after reductions in alcohol consumption and problems in Ontario 1975-1983. *British Journal of Addiction* 85: 463-468.

Smart, R. G. and Mann, R. E. (1995). Treatment, health promotion and alcohol controls and the decrease of alcohol consumption and problems in Ontario: 1975-1993. *Alcohol and Alcoholism* 30: 337-343.

Smart, R. G., Mann, R. E. and Anglin, L. (1989). Decreases in alcohol problems and increased Alcoholics Anonymous membership. *British Journal of Addiction* 84(5): 507-514.

Smart, R. G. and Ogborne, A. C. (1994). Street youth in substance abuse treatment: Characteristics and treatment compliance. *Adolescence* 29: 733-745.

Smart, R. G. and Walsh, G. W. (1995). Do some types of alcoholic beverages lead to more problems for adolescents? *Journal of Studies on Alcohol* 56: 35-38.

Sobell, L. C., Sobell, M. B., Riley, D. M., Klajner, F. and Leo, G. (1986). The effect of television programming and advertising on alcohol consumption in normal drinkers. *Journal of Studies on Alcohol* 47(4): 333-340.

Sobell, L. C., Sobell, M. B., Toneatto, T. and Leo, G. I. (1993). What triggers the resolution of alcohol problems without treatment? *Alcoholism: Clinical and Experimental Research* 17(2): 217-224.

Sommer, R. (1965). The isolated drinker in the Edmonton beer parlor. *Quarterly Journal of Studies on Alcohol* 26: 95-110.

Sons of Temperance (1876). *Sons of Temperance Blue Book*. Toronto: Sons of Temperance.

Spence, B. H. (1925). *Quebec and the Liquor Problem*. Westerville, OH: American Issue Publishing Co.

Spence, R. E. (1919). *Prohibition in Canada*. Toronto: Dominion Alliance, Ontario Branch.

Spivak, K., Sanchez-Craig, M. and Davila, R. (1994). Assisting problem drinkers to change on their own: Effect of specific and non-specific advice. *Addiction* 89: 1135-1142.

Statistics Canada (1989). *Aboriginal People Output Program: Census Canada 1986*. Ottawa: Ministry of Supply and Services Canada.

Statistics Canada (1993). *Language, Tradition, Health Lifestyle and Social Issues: 1991 Aboriginal Peoples Survey*. Ottawa: Statistics Canada, Catalogue #89-533.

Stewart, K. and Klitzner, M. (1990). Youth anti-drinking driving programs. In *Drinking and Driving*, ed. Wilson, R. J. and Mann, R. E. New York: Guilford Press.

Storm, T and Cutler, R. E. (1981). Observations of drinking in natural settings: Vancouver beer parlors and cocktail lounges. *Journal of Studies on Alcohol* 42: 972-997.

Strickland, S. (1970). *Twenty-seven Years in Canada West or the Experience of an Early Settler*. 1853 Reprint, Edmonton: Hurtig Publishers.

Sundby, P. (1967). *Alcoholism and Mortality*. Oslo: National Institute for Alcohol Research, Universitete-forlaget.

Surion, J. L. (1966). The economic effects of state monopoly of packaged liquor retailing. *Journal of Political Economy* April, 188-194.

Tawney, R. H. (1929). *Religion and the Rise of Capitalism, a Historical Study*. London, J. Murray.

The Canadian Musical Fountain: For Temperance Meetings, Bands of Hope, Temperance Conventions, Social Gatherings, Home Circles, etc. (1874). Toronto: J. Campbell.

Thompson, J H. (1969). *The Prohibition Question in Manitoba, 1892-1928*. Master's thesis, University of Manitoba.

Tu, G.-C. and Israel, Y. (1995). Alcohol consumption by Orientals in North America is predicted largely by a single gene. *Behavior Genetics* 25: 59-65.

Tyrell, I. (1990). *Women's World: Women's Empire*. Chapel Hill, NC: University of North Carolina Press.

U.S. Department of Health and Human Services (1993). *Eighth Special Report to Congress on Alcohol and Health*. Washington: DHSS.

Valle, S. K. (1981). Interpersonal functioning of alcoholism counselors and treatment outcome. *Journal of Studies on Alcohol* 42(9): 78-90.

Vingilis, E., Blefgen, H., Lei, H., Sykora, K. and Mann, R. (1988). An evaluation of the deterrent impact of Ontario's 12-hour license suspension law. *Accident Analysis and Prevention* 20: 9-17.

Vingilis, E. and Salutin, L. (1980). A prevention program for drinking and driving. *Accident Analysis and Prevention* 12: 267-274.

Vingilis, E. and Smart, R. G. (1981). Effects of raising the legal drinking age in Ontario. *British Journal of Addiction* 76: 415-424.

Volpicelli, J. R., Clay, L. L., Watson, H. T. and Volpicelli, L. A. (1994). Naltrexone and the treatment of alcohol dependence. *Alcohol Health and Research World* 18: 272-278.

Walsh, D. C., Hingson, R. W., Merrigan, D. M., Levenson, S. M. et al. (1991). A randomized trial of treatment options for alcohol-abusing workers. *New England Journal of Medicine* 325: 775-778.

Walsh, M. E. and McLeod, D. A. D. (1983). Breath alcohol analysis in the emergency department. *Injury* 15: 62-66.

Warsh, C. K. (1989). *Moments of Unreason: The Practice of Canadian Psychiatry and the Homewood Retreat, 1883-1923*. Kingston: McGill-Queen's University Press.

Warsh, C. K. (1993). *Drink in Canada: Historical Essays*. Montreal and Kingston: McGill-Queen's University Press.

Weisner, C. (1987). The social ecology of alcohol treatment in the United States. In *Recent Developments in Alcoholism*, Vol. 5., ed. Galanter, M. New York: Plenum Press, 203-243.

West, P., Dick, R., Ferris, J., Giesbrecht, N. and Wong, S. (1994). *The Maple Leaf Gardens Study*. Toronto: Addiction Research Foundation.

West, P., Giesbrecht, N. and Wong, S. (1994). Influences on helping others and the intent to reduce drinking: Drinking patterns, alcohol-related problems, and demographic characteristics. Paper presented at a conference on Addiction and Mutual Help Movements in a Comparative Perspective. Toronto, September 1994.

Whitehead, P. C. (1975). The prevention of alcoholism: Divergences and convergences of two approaches. *Addictive Diseases* 1: 431-443.

Whitehead, P. C. (1977). Deterrence of drinking-driving: Priorities in public policy. *Canadian Journal of Public Health* 68: 447-451.

Whitehead, P. C. (1983). *Young Drinkers: Recent Epidemiology.* Ottawa: Health Promotion Branch, Health and Welfare Canada.

Whitehead, P. C. and Simpkins, J. (1983). Occupational factors in alcoholism: In *Research Advances in Alcohol and Drug Problems,* Vol. 5. Ed. Kalant, O. J. New York: Plenum Press.

Whitehead, P. C. and Szandorowska, B. (1977). Introduction of low-alcohol content beer: Test of addition-substitution hypothesis. *Journal of Studies on Alcohol* 38: 2157-2164.

Whyte, J. M. (1898). *Nuggets of Gold for Temperance Campaigns.* Toronto: W. Briggs.

Willenbring, M. L., Johnson, S. B. and Tan, E. (1994). Characteristics of male medical patients referred for alcoholism treatment. *Journal of Substance Abuse Treatment* 11(3): 259-265.

Williams, M. (1984). Alcohol and the elderly: an overview. *Alcohol Health and Research World* 8: 3-9.

Wilsnack, S. C., Wilsnack, R. W. and Klassen, A. D. (1986). Epidemiological research on women's drinking 1978-1984. In *Women and Alcohol: Health-related Issues.* Washington, DC: US Government Printing Office, Research Monograph #16, DHSS Pub. No. ADM86-1139: 1-68.

Women's Christian Temperance Union (1994). *Annual Report,* WCTU Toronto.

World Health Organization (1992). *The ICD-1-Classification of Mental and Behavioral Disorders, Clinical Descriptions and Diagnostic Guidelines.* Geneva: WHO.

Zwarun, S. (1978). The (well-scrubbed) face on the barroom floor. *Maclean's,* March 1978, 62-63.

Zweben, A., Pearlman, S. and Li, S. (1988). A comparison of brief advice and conjoint therapy in the treatment of alcohol abuse: The results of the marital systems study. *British Journal of Addiction* 83: 899-916.